RUSSIAN HOMINOLOGY
THE BAYANOV PAPERS

The founders of hominoid research in Russia: (Left to right) Boris Porshnev, Alexander Mashkovtsev, Pyotr Smolin, Dmitri Bayanov, and Marie-Jeanne Koffmann. The photograph was taken in January 1968. Boris Porshnev arranged a photographer, invited his very close friends and colleagues and said, "This is for us to be remembered in the future."

RUSSIAN HOMINOLOGY
THE BAYANOV PAPERS

Dmitri Bayanov

Edited by:
Christopher L. Murphy
and
Todd Prescott

ISBN-13: 978-0-88839-736-2
ISBN-10: 0-88839-736-4
Copyright © 2014 Dmitri Bayanov

Library and Archives Canada Catalogue in Publication
Bayanov, Dmitri
Russian Hominology: The Bayanov Papers.
Includes bibliographical references and index.

All rights reserved. No part of this publication may be reproduced, stored in a retrieval system or transmitted, in any form or by any means, electronic, mechanical, photocopying, recording, or otherwise, without the prior written permission of Hancock House Publishers.

Editors: Christopher L. Murphy and Todd Prescott
Book design: Christopher L. Murphy
Front Cover photo: D. Bayanov; **Cover design:** Ingrid Luters
Back cover photos: Top: painting titled *Fauni Danzanti* by Hans Thoma, 1919; Center: Mosaic showing Mask of Silenus (a satyr), Roman 1st century; Lower: A domovoy – house spirit in Slavic folklore. All photos public domain.

We acknowledge the financial support of the Government of Canada through the Canada Book Fund for our publishing activities.

Published simultaneously in Canada and the United States by:

HANCOCK HOUSE PUBLISHERS LTD.
19313 Zero Avenue, Surrey, B.C. Canada V3S 9R9
(604) 538-1114 Fax (604) 538-2262
HANCOCK HOUSE PUBLISHERS
#104-4550 Birch Bay Lynden Rd, Blaine, WA U.S.A. 98230
(604) 538-1114 Fax (604) 538-2262
Website: www.hancockhouse.com
Email: sales@hancockhouse.com

Contents

Acknowledgments 6
Introduction 7

Chapters:

1. Historical Evidence for the Existence of Relict Hominoids 10
2. Learning from Folklore 43
3. Hominology in the Balkan Peninsula 113
4. Wheatcroft's Orang Pendek Evidence – Thoughts 128
5. The Harm of Assumptions Turned into Convictions 140
6. Thoughts on the Revolution in Anthropology 143
7. The Problem of Acknowledgement of Hominology by the Scientific Community 177
8. Dr. Koffmann Replies to Professor Avdeyev 181
9. Brief Ecological Description of the Caucasus Relic Hominoid (almasty) by Dr. Marie-Jeanne Koffmann 184

Final Concluding Remarks 195

Color Photo Presentation 160
Bibliography 196
General Index 200

Notes on word terms and usage: Throughout this work several variations of the root word "homin" (or this word alone) have been used. All such words generally mean, "of or related to sasquatch/bigfoot, almasty or associated beings." The word "hominology" refers to the study of such beings. Also, it should be noted that for the purpose of this book the word "almasty" has been used to identify the Russian equivalent of sasquatch/bigfoot.

Furthermore, the words sasquatch, bigfoot, yeti, almasty, yowie, and yeren have not been spelled with a capital letter except in quoted material. Also, the word "sasquatch" is used as both a singular and plural word.

Special Notes: As this volume is a collection of papers (essays, articles, talks, and so forth), in some cases information, supporting stories or references, have been repeated in different papers. Nevertheless, to avoid excessive repetition, I have referred the reader to the first paper that contains the applicable information when more than a paragraph is involved.

It also needs to be noted that editing of all material was necessary for this volume. As a result, some material will differ from that seen in my original papers or articles.

Acknowledgments

Special thanks is extended to Kathy Moskowitz Strain for the use of her material, and the memory of the late Professor Boris Porshnev for the knowledge he imparted to me.

I also wish to acknowledge the late Bobbie Short who kindly posted my articles to her website, and now Candy Michlosky who has provided them on her website.

I am grateful to publisher David Hancock for his interest in my works and publishing them. But the greatest thanks and appreciation go to my editor Chris Murphy for his huge work in preparing this volume and my previous one. He is the best editor I've had in my life and that is largely because he himself is a writer—author of most valuable hominology books. So there is no lack of mutual understanding. As a rule, he accepts my proposals and I accept his. Our author-editor relations are ideal.

In the final preparation of this work Chris asked Todd Prescott to assist him with detailed editing and proofing. Todd kindly agreed and I extend my appreciation to him for this great favor.

Thanks also to Wikipedia for the information I obtained from this source.

Dmitri Bayanov

Introduction

In 1964 I met Professor Boris Porshnev, read his book *The Present State of the Question of Relict Hominoids* (416 pages—only 180 copies printed by the ruling of the Soviet Academy of Sciences), and became a participant in the research, called since 1972, *Hominology*. My first Caucasus expedition, led by Marie-Jeanne Koffmann, was a great eye-opening event in my life. I realized then and there that Porshnev was right in saying that under the mythological names of *devils, wood goblins, domovoys (brownies)*, etc., stood *real beings!* That transpired from the fact that local witnesses of hairy "wildmen" called them interchangeably by corresponding ethnic names, such as *almasty, kaptar, meshi adam*, etc., as well as *devils, shaitans, wood goblins*, etc.

I knew, as all people do, that in popular fairy-tales the names of animals—bears, wolves, foxes, and so forth—indicate imaginary mythological creatures, whose real counterparts exist in nature. In a Russian folk tale a bear walks on a wooden leg, and in another tale the cunning fox is riding on the back of the simpleton wolf. So the names of real beings are also used for mythological entities. To my great surprise I learned that people, living in much closer contact with nature than I was, used mythological names, such as *wood goblin*, for indicating hairy "wild men" who they regarded as real. A local man said to Koffmann: "There are wild goats, wild rams, wild hogs. Why shouldn't there be wild men?"

Indeed, why? Ah, simply because peasants call such hairy wildmen by the names "goblins," "devils," "shaitans" etc. For educated people, for men of science especially, this means nothing but mythology. The main argument of academic opposition to Porshnev was that he took popular myths, the wood goblin myths in particular, for reality. Incidentally, academic opposition to the reality of bigfoot/sasquatch in North America is deadbent on using the same argument.

This set me to study folkloristics, demonology and the history of religion in order to fill in this shameful gap between the knowledge of common people and the ignorance of scholars. Soon I came up with the work *In Defense of Devilry,* claiming that there was a reality to devils, shaitans and wood goblins. The work couldn't be published in the Soviet Union with its restrictive and dogmatic ideology. It was published in 1991, the year of the Soviet Union's disintegration; the book's title changed to *Wood Goblin Dubbed Monkey: A Comparative Study in Demonology*. The work was based on the study of the ethnic folklore and demonology of many peoples of the former Soviet Union.

Among Professor Porshnev's many opponents, the toughest and fiercest was zoologist and paleontologist Professor Nikolai Vereshchagin (1908–2008), who called our research "pseudoscience," and was most sarcastic and critical in his article, "Wood Goblins of the 20th Century." I wrote about the battle waged by Vereshchagin and his colleagues against Porshnev and hominology in *America's Bigfoot: Fact, Not Fiction* (pages 100-106). So for laugh's sake, I sent him a copy of my demonology book, inscribed "With Greetings from the Wood Goblins of the 20th Century," but never expected a response. To my surprise it did come, with the opening words, "Dear Dmitri Yurievich [my patronymic name], I received your excellent book about devils the other day," and ending with, "Thanks for the book. I wish you success." But in the middle of the message he bad-mouthed Porshnev again, calling him a "paranoiac." As a result, it took me some time to explain to the professor that without Porshnev my "excellent book" could not have appeared, and that I wished the science world would have as many "paranoiacs" like Porshnev as possible. To make a long story short, we became friends and he stopped denying the reality of wood goblins, but did not become our ally in practice because of his age. He died in 2008, one month short of age 100.

I take this conversion of the worst critic into a friend and supporter after reading my book and communicating with me as my major achievement in hominology. It shows that we could win over many scientists, who are critical or don't care about our research, if we could let them know the truth we possess instead of the misinformation they have from the tabloids and the rest of the mass media—as well as direct lies in the books of dedicated debunkers. This point pertains in particular to this book because folklore makes up a large part of it.

In 2009, U.S. anthropologist Kathy Moskowitz Strain kindly presented me with her large book *Giants, Cannibals & Monsters: Bigfoot in Native Culture*, which is a treasure-trove of North

American Native folklore on what we call bigfoot and sasquatch. Naturally, I couldn't help examining it and learning from it the way I examined and learned from the similar folklore in Europe and Asia. The latter is presented in Chapter 1– Historical Evidence for the Existence of Relict Hominoids (a paper written for The Relict Hominoid Inquiry Internet site); the former in Chapter 2 – Learning from Folklore, a paper that along with others (Chapters 3 to 7) did not make their way into my book *Bigfoot Research: The Russian Vision*. The last part is devoted to material by Marie-Jeanne Koffmann whose exchanges with Professor Valeri Avdeyev in the 1960s press sound very topical today. Her paper on the ecology of almasty presents the strictly factual aspect of hominology equally dealt with in this book.

Professor Porshnev believed that our research would bring a revolution in science. I hope to live to witness it.

D.B.
Moscow, November 2013

CHAPTER 1

Historical Evidence for the Existence of Relict Hominoids

(Published on *The Relict Hominoid Inquiry* website, Idaho State University, 2012.)

ABSTRACT: Hominology is the study of evidence for the existence of wild bipedal primates, presumed to be relict hominoids or hominids. Investigation of the subject began simultaneously in Russia and America last century, beginning with the Himalayan expeditions in search of the yeti. The first international scientific organization that united academic and non-academic investigators was formed and functioned in Italy in the 1960s. Its Russian member was Dr. Boris Porshnev, founder of Russian hominology, whose unorthodox views regarding the origin of man and the nature of hominids are pointed out. Hominology is based on six main categories of evidence, of which two, pertaining to the historical aspect of the subject, are discussed in detail in this essay. They are the evidence of natural history, from Lucretius to Linnaeus, and the evidence of myth and folklore, from Babylonian mythos to folk proverbs and sayings in use today. The reinforcement of early natural historians' descriptions by cultural literary traditions attests to the acceptance of wildmen, a.k.a. demons, devils, goblins, as hair-covered creatures in human form. In the author's view, present data testify that hominology deals with evidence of living pre-sapiens relict hominoids.

INTRODUCTION: Systematic hominology in Russia and North America has many similarities and certain differences. In both regions it began in the middle of last century, stimulated by the Himalayan expeditions in search of the yeti. The founders of the research were Bernard Heuvelmans, Ivan Sanderson and Boris Porshnev. They agreed on one thing—that wild, hairy bipeds are real. However, they disagreed on almost everything else. Heuvelmans and Sanderson were zoologists; Porshnev was a historian and philosopher versed in many scientific disciplines. For

Heuvelmans and Sanderson the problem was zoological; for Porshnev it was above all anthropological, pertaining to the origin and position of man (Fig. 1). His theory of man's origin was different from that of mainstream anthropologists, and he held that the evidence for the existence of wild bipedal primates perfectly matched and supported his theory. The theory's thesis being that speech and its morphological and neurological correlates are the species-specific characteristics of *Homo sapiens*. He maintained that all pre-sapiens bipedal primates, including Neanderthals, were devoid of the faculty of speech, and therefore belonged to the animal kingdom. In this connection he proposed to change the term for the family *Hominidae* to *Troglodytidae*, and he believed that the extant wild hairy bipeds, reported today, were relicts of Neanderthals, who stopped making and using stone tools and fire (or lost these skills to a significant degree) due to a greatly changed environment, dominated by *Homo sapiens*. It should be noted that recent review of archeological evidence raises questions of whether Neanderthals were habitual fire-users during the Mousterian, and indicates that it may be possible that fire use was not a significant component of the Neanderthals' adaptation to their local environments. (Sandgathe, et al., 2011). The origin of *Homo sapiens* is thus viewed as tantamount to the origin of speech (Porshnev, 1974; Bayanov and Burtsev, 1974, 1976).

Figure 1. Boris Fedorovich Porshnev (1905–1972), the founder of Russian hominology. (Photo: D. Bayanov)

Porshnev, Sanderson, and Heuvelmans were good friends and members of The International Committee for the Study of Hairy Humanoids (the name owes its origin to Heuvelmans), an organization created in Rome in 1962 by Dr. Corrado Gini, Emeritus Professor of Sociology at Rome University. Opening the Committee, Dr. Gini said, in full agreement with Boris Porshnev,

"The Snowman and other hairy bipeds present a subject worthy of a profound scientific study. (...) This is a subject of the greatest importance for understanding the origin of man and the initial stages of human society." (*Genus*, 1962).

The Committee included some 30 persons from different countries, among them Dr. George Agogino, Dr. Raymond A. Dart, Dr. John Napier, Dr. W. C. Osman Hill, Dr. P .R. Rinchen, Prof. Philip V. Tobias, as well as yeti investigator Ralph Izzard, yeti and bigfoot investigators Tom Slick and Peter Byrne, sasquatch investigators John Green, Bob Titmus, and René Dahinden.

The journal *Genus* (not peer-reviewed), published by Gini, printed many articles by the Committee members, e.g., "Almas still exists in Mongolia," by P. R. Rinchen; "Report on a Sample of Skin and Hair from the Khumjung Yeti Scalp," by M. Burns, "Being Some Notes, in Brief, on the General Findings in Connection with the California Bigfoot," by Peter Byrne, "Hairy Primitives or Relic Submen in South America," and "Preliminary Description of the External Morphology of What Appeared to be the Fresh Corpse of a Hitherto Unknown Form of Living Hominid," (so-called Minnesota Iceman – D.B.) by Ivan T. Sanderson, as well as a number of articles in French, Italian, and Spanish, contributed by, among others, Porshnev, Gini, and Heuvelmans.

The organization ceased to function after the death in 1965 of its creator. Had it continued to exist, I am sure our situation today would be quite different, as the Committee included prominent academics who provided a vital link with mainstream science. After a break of forty-five years this favorable condition is being revived and re-established anew with the creation of The Relict Hominoid Inquiry.

HOMINOLOGY: Boris Porshnev envisaged our research as a new and distinct discipline, which I named "hominology." Not surprisingly, terminology for the objects of hominology proved a protracted problem. Porshnev used the term relict hominoid, actually implying relict hominid in the classification generally accepted at the time. I have used both terms interchangeably, always implying "hominid." For the sake of convenience, by way of *professional* jargon, I have also been using a contraction—"homin"—as a substitute for hominoid, hominid, wild bipedal primate, wild man, yeti, almasty, sasquatch, and the rest of ethnic names for the creatures under study. This term also serves to avoid the current state of transition in the substitution of hominin for hominid in the technical literature, in accordance with the current cladistic approach to taxonomy.

Hominology's database consists of the following main categories:

1. Natural history
2. Folklore and mythology
3. Ancient and medieval art
4. Eyewitness testimony
5. Footprint evidence
6. Photographic evidence

In this essay I will limit my treatment to the first two areas of the historical aspects of hominology in the Old World, using as illustrations samples of ancient and medieval art from the third category. Today, a corresponding collection includes scores of hominid images (pictures, sculptures, petroglyphs) from across the world. It presents two kinds of portrayal: realistic and "ritualistic," i.e., symbolic. The first is true to life and helps the hominologist to study the creatures' appearance and anatomy. They show hairy bipeds with certain typical features setting them apart from humans. Symbolic portrayals may be a caricature that shows not so much the real object as the artist's attitude towards it. Images of grotesque monsters in ancient and medieval art have therefore led scientists and art specialists to believe that these monsters were merely figments of the imagination, with no basis in reality. Hominology offers a potential alternative to such views.

Natural History: A celebrated source here is Lucretius Carus (1st century BC), who in his famous *De rerum natura* (On the Nature of Things) describes a race of wildmen who had very strong bodies covered with hair. These wildmen lived in woodlands and caves with neither language nor clothes or industry. They hunted animals with sticks and stones and ate meat and other foods raw. It is most remarkable that Lucretius says that these woodland wildmen were ancestral to modern man (Lucretius, 1947).

Greco-Roman naturalists used the word troglodyte (caveman) to denote bipeds that were different from humans. Among the emphasized characteristics of troglodytes were the creatures' great speed in running, lack of intelligible speech, and strange vocalizations (Pliny, 1979: 5, 8). Popular names in the Greco-Roman world for these creatures were *satyr, silenus, faun,* and *pan* (Fig. 2). Roman naturalist Pliny the Elder (1st century AD) says in his *Natural History* that "the Satyrs have nothing of ordinary humanity about them except human shape." (Pliny, 1979).

Figure 2. As nimbuses serve to identify divine persons in Christian art, so horns, hoofs and tails indicate heathen gods of hominoid origin in ancient art. Here is an image of the ancient Greek god Pan, patron of herdsmen, hunters (circa. BC 100). (Photo: Public domain)

Geographer Pausanias (2nd century AD), in his Description of Greece, says "That the Silenuses are a mortal race may be inferred especially from their graves; for there is a tomb of one Silenus in the land of the Hebrew, and there is the tomb of another at Pergamus." (Pausanias, 1913: VI, XXIV). We also learn from him the following: "Elderly Satyrs are named Silenuses (Fig. 3). Wishing to know particularly who the Satyrs are, I have for that purpose talked with many persons." This shows that already in ancient Greece the creatures in question were considered enigmatic. The author continues: "Euphemus, a Carian, said that when he was sailing to Italy he was driven by gales out of his course and into the outer ocean, into which mariners do not sail. And he said that there were many desert islands, but that on other islands there dwelt **wildmen** (my emphasis – D.B.). The sailors were loath to put into these latter islands... These islands, said he, are called by the seamen the **Isles of the Satyrs**" (Pausanias, 1913: I, XXIII). The identification of **satyrs** with **wildmen** is noteworthy.

The enigmatic nature of satyrs at the time is also confirmed by

Plutarch, who tells of an actual capture of a satyr by the soldiers of the Roman general Sulla in the territory of modern Albania, in the year 86 BC. The satyr was brought to Sulla and "interrogated in many languages as to who he was; but he uttered nothing intelligible; his accent being harsh and inarticulate, something between the neighing of a horse and the bleating of a goat." The general "was shocked with his appearance and ordered him to be taken out of his presence." (Plutarch, 1792: 349).

Figure 3. Sculpted portryal of silenus found in the excavation of Nymphaion, an ancient Greek colony in the Crimea. The term "silenus" denoted an "old satyr." (Photo: Public domain)

From the Middle Ages an important piece of information comes from the Persian scholar Nizami al-'Arudi (12th century AD). In his book *Chahar Maqala,* he says that the lowest animal is the worm and highest is Nasnas, "a creature inhabiting the plains of Turkistan… This, after mankind, is the highest animal, in as much as in several respects it resembles man: first in its erect stature, secondly in the breadth of its nails, and thirdly in the hair on its head." (Bernheimer, 1952: 190).

Interesting information comes from medieval Arab travelers who visited the Caucasus in the 10th century AD and wrote that the forests there "are inhabited by a sort of monkey having an erect stature and round face; they are exceedingly like men, but they are all covered with hair… They are deprived of speech… They express themselves by signs." The Arab author, Abul Hasan Ali Masudi, also mentions the existence of "monkeys that approach in appearance the figure of man" in the land of the Slavs and other nations in the territory of modern Russia (Masudi, 1841: 440).

In the 15th century, a native of Bavaria, Johann Schiltberger, was taken prisoner by the Turks and sold to the Khan of Siberia. After 30 years spent in Asia, Schiltberger returned home to Bavaria, and in his book of travels described "savages, who are not like other people... They are covered all over their body with hair, except the hands and face, and run about like other wild beasts in the mountains, and also eat leaves and grass and any thing they can find. The lord of the country sent to Edigei [another ruler of the land – D.B.] a man and a woman from among these savages, that had been taken in the mountains" (Schiltberger, 1879: 35).

In Europe, Albertus Magnus (1193-1280), a philosopher deeply interested in natural science, narrates in his *De Animalibus* (2.1.4. 49–50) of the recent capture in Saxony of two (male and female) forest-dwelling, hairy monsters much resembling human beings in shape. The female died of blood poisoning caused by dog bites, while the male lived on in captivity and even learned the use of, albeit very imperfectly, a few words.

Figure 4. A portrayal of the hominoid side by side with *Homo sapiens* in the 13th century sculpture of a peasant and a wild man on the north portal of Notre Dame, Semur-en-Auxois, Burgundy, France. (Photo: Public domain)

One of the most realistic portrayals of the hominoid side by side with *Homo sapiens* is the 13th century sculpture of a peasant and a wild man on the north portal of Notre Dame, Semur-en-Auxois, Burgundy, France (Fig. 4). The low cranial vault, prominent brows, large orbits and prominent cheek bones, receding chin, and how the head is set on the shoulders all bespeak a typical Neanderthal.

The tradition of the wild man's presence in medieval Europe is well documented by U.S. scholar Richard Bernheimer (1952) in his book, *Wild Men in the Middle Ages*. Every aspect of the theme is covered and discussed on the basis of historical documents and works of art in the following chapters:

1. The Natural History of the Wild Man
2. His Mythological Personality
3. His Theatrical Embodiment
4. The Learned Aspect
5. The Erotic Connotation
6. His Heraldic Role

Here are some important quotes from Chapter 1:

About the wild man's habitat and manner of life, medieval authorities are articulate and communicative. It was agreed that he shunned human contact, settling, if possible, in the most remote and inaccessible parts of the forest, and making his bed in crevices, caves, or the deep shadow of overhanging branches. In this remote and lonely sylvan home he eked out a living without benefit of metallurgy or even the simplest agricultural lore, reduced to the plain fare of berries and acorns or the raw flesh of animals. (Bernheimer, 1952: 9).

Medieval writers are fond of the story which tells how hunters, venturing farther than usual into unknown parts of the forest, would chance upon the wild man's den and stir him up; and how, astounded at the human semblance of the beast, they would exert themselves to capture it, and would drag it to the local castle as a curiosity... The wild man's own reaction to the sudden encounter with his civilized counterpart varies according to type and temperament. (...) But whether they be elusive or combative, the result of the encounter is the same: the wild man is dragged out of his habitat and brought to the castle, there confined, and immediately exposed to the efforts of his captors to return him to full-fledged human status. Only if all endeavor fails, and the hairy man remains morose and speechless in spite of blandishment or torture, can he hope to be released again. (Bernheimer, 1952: 17).

The wild man holds thus a curiously ambiguous and ill-defined position in God's creation, being neither quite man enough to command universal agreement as to his human identity, nor animal enough to be unanimously classified as such. (Bernheimer, 1952: 6).

In many ways his life resembled that which we now attribute to the raw beginnings of human cultural existence in the Stone Age." (Bernheimer, 1952: 10).

After reading the above, one can think that the author is what I call a perfect "realist," and not a "folklorist," regarding the existence of "wild men." But that is not so, as is clear from the very first page of the book:

> Since the title of this book is startling, implying a concern with madness, passion, and violence, it may be well to assure the reader from the start that wild men are **imaginary creatures** [my emphasis – D.B.] and that their name is a technical term. It would be difficult, in fact, to find another less shocking name for them, since the one employed here has been in common usage ever since the Middle Ages and is one of the few which denote the subject unambiguously. This book does not deal with actual outlaws, lechers, and bad men then or at least not primarily. Instead it deals with a literary and artistic figure whose imaginary character is proved by its appearance: **it is a hairy man curiously compounded of human and animal traits, without, however, sinking to the level of an ape.** [my emphasis – D.B.]

This makes me wonder how the author may have reacted to the description of bigfoot/sasquatch. Probably in the usual way of his peers, as follows from his words:

> It appears that the notion of the wild man must respond and be due to a persistent psychological urge. We may define this urge as the need to give external expression and symbolically valid form to the impulses of reckless physical self-assertion which are hidden in all of us, but are normally kept under control." (p. 3)

One of the most detailed and trustworthy accounts of a European wild man in captivity was published in Vienna in 1796, by Michael Wagner, in his scholarly *Beiträge zur philosophischen Anthropologie*. It dealt with a hairy wild man of perfect Neanderthal anatomy, captured in Rumania and held in captivity in the city of Kronstadt (now Brasov) in the second half of the 18th century (Wagner, 1796; Singh and Zingg, 1942).

It is a curious fact of anthropology that its basic term—*Homo sapiens*—owes its origin to the existence of troglodytes. It is generally believed that the term was coined to distinguish modern man from extinct forms known from the fossil record. That is not so. The term *Homo sapiens* was introduced by Linnaeus in the middle of the 18th century, a hundred years before Darwinian theory and knowledge of, let alone systematic studies of hominid fossils. Linnaeus had information from Pliny the Elder and other ancient authors, as well as from contemporary Dutch explorers in Southeast Asia—Bontius, Rumphius, etc.,—about the existence of man-like bipedal primates, hairy, speechless, non-sapient, and for the sake of contrast with them he designated our own species with the rather wishful term *"sapiens"* (the wise) (Linnaeus, 1758, 1760).

It was with awe that one day, in 1966, I opened and copied relevant pages in Latin from the original 10th edition of Caroli Linnaei's *Systema Naturae* (1758), in the library of the Moscow Zoological Museum. This edition launched the Linnaean nomenclature. One of its salient features is that it presents two living species of man: *Homo sapiens* (man the wise) and *Homo troglodytes* (caveman). The first is described as "diurnus, varians cultura, loco," the second as "nocturnus" and "sylvestris." *Homo sapiens* is subdivided into races, and includes *Homo ferus,* which designated, in the opinion of Linnaeus, *Homo sapiens* gone wild (children captured and reared by animals), but actually embraced also some cases, as it is apparent now, of real "wild men" (i.e., relict hominids) reported at the time in Europe. Right after the term Homo sapiens, Linnaeus put in the words to address mankind, *"Nosce te ipsum"* (know thyself).

Homo ferus and *Homo troglodytes* evidently filled in for Linnaeus the gap between ape and man and prompted him to establish a single Order of Primates. On the one hand, there were human children reared by animals and turned into beasts; on the other hand, stood *Homo troglodytes* that seemed to be more man-like than ape-like, especially on account of bipedalism and the dental system devoid of diastemata, the characteristic of apes and monkeys. (His information included this important detail). So there is no doubt that man owes his undeserved name of *Homo sapiens* to the presence of non-sapient *Homo troglodytes* in the Linnaean classification.

Still, his information on the subject was so patchy, fragmentary, and contradictory that the great classifier, with his passion for order and exactness, must have been tormented by the lack of precise knowledge in the matter. This is seen from the dissertation *Anthropomorpha* (Fig. 5), which he dictated (as was the custom at the time) to his St. Petersburg student, Christian Hoppius, saying in part the following:

Figure 5. *Troglodyta Bontii*, alias *Homo sylvestris*, witnessed and depicted by Jacobus Bontius in Java in the 17th century (published in 1658). (Photo: Public domain)

Is it not amazing that man, endowed by nature with curiosity, has left the troglodytes in the dark and did not want to investigate the creatures that resemble him to such a high degree? A lot of mortals spend their days in feasts and banquets, and all they care for is how to prosper by honest and dishonest means. No better is the behavior of most navigators who sail to the Indies and who alone happen to see the troglodytes. Driven by greed, they despise the tasks of natural science, such as investigation of the way of life of troglodytes. Just imagine what wondrous objects of diversion for a monarch in his palace such animals could be, for one would never tire of marveling at them. Or is it really difficult for a monarch to get such animals, knowing that people vie with each other to fulfill his orders? And it would be of no small benefit for a philosopher to spend several days in the company of such an animal in order to investigate how much superior human reason is and thus discover the difference between those endowed with speech and those devoid of it. And should I mention what light could be shed for natural science from a detailed description of these animals. As for me, I remain in doubt what specific characteristic distinguishes the troglodyte from man within the scope of natural history.

(Linnaeus, 1760). *[My translation from a Russian translation from the Latin, published in St. Petersburg in 1777. The original Latin text appears to be lost. – D.B.]*

The fervent call of the great naturalist fell on deaf ears. Not only that, but his whole classification of primates, along with the latter novel term (introduced by him in zoology), was condemned and done away with by the scientific establishment of the century, whose faith revolted against Linnaeus's innovations. The job was done by Johann Blumenbach, who in his *Manual of the Natural History* (1775) established the order *Bimanus* for man and the order *Quadrumanus* for apes and monkeys. As for *Homo troglodytes*, Blumenbach discarded the species altogether as "an unintelligible mixture of pathological cases and the orangutan." He moved the term *"troglodytes"* to *Simia* and established *"Simia troglodytes* or Chimpansi," which implied that chimps were cave-dwellers.

According to Stephen J. Gould, "Historical changes in classification are the fossilized indicators of conceptual revolutions." Blumenbach's monumental change in the Linnaean classification was then a conceptual counter-revolution, which lasted nearly a hundred years, until resisted and reversed by Darwin's "bulldog," Thomas Huxley, who with *Man's Place in Nature* (1863) restored the single order of Primates, as well as the term itself. But *Homo troglodytes* stayed in limbo for another hundred years, until resurrected and vindicated by Boris Porshnev in *The Present State of the Question of Relict Hominoids,* proclaiming yet another conceptual revolution (Porshnev, 1963).

Folklore and Mythology: During my first expedition to the Caucasus in 1964, I was struck by the fact that the locals often referred to the reported hairy wild man quite matter-of-factly by such names as "devil," "satan," "wood goblin," etc. Back in Moscow, I plunged into reading literature on folklore, demonology, and the history of religion. I was fascinated by what opened to my eyes, my mind already opened by the Porshnev theory and what I learned during the expedition.

It became clear to me that folklore and demonology, or what Dr. John Napier called the "Goblin Universe," is a rich source of hominology, quite realistic, but largely misunderstood and misinterpreted by academic specialists on folklore and mythology. Soon I came up with a work whose title could be translated into English as *In Defense of Devilry*. The work was never published in Soviet years and no folklorist ever agreed to collaborate with me.

When the country's political situation began to change, I enlarged my original work, changed the title to *Wood Goblin Dubbed Monkey: A Comparative Study in Demonology*, and after addressing in vain many publishers, at last succeeded in finding one who published it in 1991. I sorted out in it volumes of published folklore of many peoples in the Soviet Union, focusing on the most realistic descriptions of the appearance, behavior, and habits of their "demons."

Academic folklorists and demonologists refer to the heroes of their books, i.e., "devils," "goblins," "brownies," etc., by such names as "fabulous beings," "creatures of fantasy," "irreal characters," "mental constructions," etc. Accordingly, they focus attention on the fabulous and imaginary. In this respect the hominologist's objective is the opposite of theirs. To get at goblin biology and ethology he has to amass and sort out as much folklore material as possible, from as many lands and regions as possible, taking into account first and foremost not what folklorists say, but what their folk informants relate. That is why it has to be a comparative study. Folklore not only supports what we learn from contemporary eyewitnesses, but provides details and particulars gone unnoticed, because folklore contains knowledge amassed and compressed over hundreds of years.

In Theodore Roosevelt's book, *The Wilderness Hunter* (1893), Roosevelt's Native companion did not want to go into a certain area for fear of the "devils" there. Roosevelt called them "forest hobgoblins." The parallels evidenced in the ethnic "demonology" of Russia and America, provide further opportunities for hominology. What follows is a brief synopsis of information presented in my book in Russian on folklore and demonology (Bayanov, 1991).

Relict hominoids (alias homins) are different from all other cryptids (objects of cryptozoology) not only in anatomy and behavior but also in the place they hold in human culture. I dare say there is no other living creature, except man himself, which figures so prominently in religion, mythology, folklore, and the arts.

We can imagine that in the hoary past, when humans were a

minority confronted by an awesome preponderance of non-human bipeds, they had no choice but to find a way of co-existence with the homins. The latter effectively dominated the environment. So humans offered a part of their hunting trophies to homins in order to placate them and be allowed to hunt and gather food in the territories occupied by the latter. As this process went on, homins became viewed as lords of nature and eventually worshiped as heathen gods. Food offerings to placate them turned into religious sacrifices.

Div (dev, dav) is a common name for the "wild man" in Persia (modern Iran) and the adjacent countries. Initially divs were worshiped like gods by heathen peoples, and this

Figure 6. A traditional Persian style illustration in which the hero lassoes and captures a div, whose image is far removed from biology into devilry. (Photo: Public domain)

explains the fact that the words in other Indo-European languages, such as "Deus" and "Divus" in Latin, "Zeus" in Greek, "divine" and "divinity" in English, are etymologically related to the word "div." It may be noteworthy that according to Greek mythology, Zeus was born in a cave.

With the advent of major religions, such as Zoroastrianism, Judaism, Christianity, and Islam, the heathen gods and their hominid prototypes were condemned and relegated to the status of demons. This dramatic process is marked by Persian rulers' relentless struggle against divs, which is vividly described by Firdausi (940-1020) in the epic *Shahnamah*, which set a standard for Persian poetry (Fig. 6). Characteristic in this respect are the following words in the epic: "Take divs for hostile creatures. They are of those who have not been blessed by God, **who have deviated from man's way** (my emphasis – D.B.), take him for a div, don't call him a man" (Korogly, 1983: 43). Thus centuries on, the notion is echoed in *The*

Oxford English Dictionary (1989, Vol. IV): "The div of ancient Persia is supposed to be the same as the European devil of the middle ages." Divs and their counterparts elsewhere were condemned not only on account of their beastly appearance, but also and mainly because of their beastly behavior. Sources of abundant relevant information in this area range from the Babylonian epic of Gilgamesh, to fairy tales, to widely used sayings and proverbs.

Quite impressive is folklore on the origin of demons. Hebrew folklore has it that God created demons on the Sabbath eve, and therefore did not have time to make them fully human (Fig. 7). But Russian peasants had a different opinion on the matter.

Figure 7. Jewish demon, a seirim (meaning of the "hairy ones") in the original Hebrew text and "devil," "satyr," even "wild goat," and "he-goat" in various Bible translations. (Photo: Public domain)

When the peasant's son inquired, "Daddy, what is meant by the devil, the leshy, the domovoy? What is the difference between them?" The father answered, "There is really no difference. They say when God created man, Satan was eager to create too, but no matter how hard he tried he could only make devils, not men. When God saw that Satan had already produced several devils, He ordered Archangel Gabriel to expel Satan and his unclean forces from heaven. Gabriel did so. The devil that fell on a wood, or forest, became the leshy (wood goblin), another, that fell on a field, became the polevoy (field goblin), and a third, that fell on a house, became the domovoy (domestic demon, brownie). That's how they came about and got different names. But actually all devils are alike." (Fig. 8).

In Bielorussia, folklorists recorded the following legend: Adam and Eve had a dozen pairs of children. When God

Figure 8. A Russian domovoy as drawn by artist Ivan Bilibin, who studied and illustrated Russian folklore. The drawing was made in 1934, long before the birth of hominology. (Photo: I. Bilibin; Public domain)

came to look at them, they showed Him six pairs, and hid the other six pairs under an oak tree. So, like we come from those six pairs shown to God, they (the demons) come from the other six pairs. Their number is the same as ours, only they are invisible because they are hidden from God.

Less civilized people, living in the lap of nature, had a different and more realistic view on the subject. Thus the Mansi, living in the taiga of Siberia, say that in making people, gods used two materials: clay and larch timber. As soon as people made of larch were produced, they dashed into the forest. Those are menkvs (wood goblins). Slow moving beings, made of clay, became ordinary people. Their lifespan is short; arms made of clay, legs made of clay, what's the use of them? If man falls into water, he drowns; if the weather is hot, water comes out of him. If men were made of larch, they would be harder and wouldn't drown in the water.

There are many other folklore versions of the theme, including the belief that demons arise from dead people who were not buried or were buried the wrong way. What is interesting and important for the hominologist, as found in such tales and legends, is people's wish to explain both great likeness and great difference between man and demon, and not the essence of the explanations, arising from fantasy and superstition.

Folklore on demons confirmed all I knew about the homin anatomy and behavior, and added things I did not know. The demonic beings are hairy manlike bipeds, sometimes bigger and always stronger than man. There are male and female demons, as well as their offspring. A shock of hair is sometimes mentioned on the heads of males, but bald-headed demons are on record as well. Females boast of long-hanging or flying head hair, sometimes disheveled, and sometimes brushed.

The Komi people in the north of Russia say their wood goblins have hair-covered ears. One folklore item in Siberia mentions hair on female breasts. The hair color ranges from black to white, with lots of browns and reds, and is likened to the fur of animals native to the particular geographic area (reindeer, bear, camel, goat, and buffalo). The attribute of hairiness is present in the local names of demons, from the Hebrew se'irim, to the medieval European pilosus, to the Russian volosatic and volosatka (literally "hairy one" for male and female). The color of the skin is swarthy, with a reddish, or yellowish, or grayish tinge. The pointed cone-shaped head is a usual feature, even reflected in the names of Russian devils and wood goblins: shishko, shishiga from shishka (cone). The eyes appear big at night when they shine "like stars." Facial features are not attractive and folklore uses the word "muzzle" in reference to a demon's protruding lower face. Lack of a neck is mentioned in one item from Siberia. Folklore dwells a lot on the enormous size of a female demon's breasts, calling them "huge" and even "frightening."

Demons in Russia are fond of tree-climbing, swinging on the branches, and diving from trees on the river bank into the water. They are excellent swimmers and divers, as well as jumpers and runners. They also love dancing and merrymaking, especially all kinds of pranks, so that Russian peasants called them "jokesters" and "pranksters." A favorite prank of rusalkas (aquatic female demons) was to catch wild geese on the river and entangle the feathers of their wings so that the birds could not fly. Or they would let the fish out of the fishermen's net and fill the latter with slime and water-plants, or divert themselves by putting out fishermen's or hunters' campfires with the water dripping from their hair covering.

One folklore item from the European part of Russia states that in olden days hunters "had to prepare gifts for the 'lord of the forest' for allowing them to hunt on his property." In later times the relationship "progressed" and an item from Siberia says that hunters there engaged in barter trade with wood goblins—the latter supply squirrels and in exchange receive generous gifts of vodka.

Folklore strongly recommends hunters not build their cabins on the forest path of wood goblins. And, custom forbids whistling in the forest and in the home so as not to alert and invite the goblins. Interestingly, I heard a similar belief regarding the "creek devil" from a local teacher, during my visit to the Yurok Indian Reservation in Northern California, in September 2003.

Folk demons also actively interact with fishermen. Not only do the homins steal from fishermen's catches, but they reportedly can also help people catch fish. According to Georgian folklore, all fish in the river are controlled by a water goblin. If a fisherman leaves food and a jug of wine on the bank and speaks nicely of the demon, he will send a lot of fish into the net.

A Mordva fisherman (in the Volga area) discovered a crying goblin child in his fishing net and let it go. Ever since then he always had good catches. Ethnic Russian fishermen would throw a bast-shoe into the water and yell, "Hey, devil, drive fish into our net!"

But the demons' greatest contractors were herdsmen. It is reported that in Russia they made secret "contracts" with wood goblins, who helped pasture the herd, find lost cows, and protect them from wolves and bears. The service was paid for with food and animals from the herd. Such deals were popular with the peasants, but kept strictly secret because they were viewed as very sinful by the Orthodox Church. It is worth mentioning that in ancient Rome fauni were said to protect herds from wolves, and a celebration was held in their honor on the 15th of February, called Lupercalia.

Another kind of interaction and category of homin I call "visiting demons" are those who approach human habitation for one reason or another. The most common reason is food, another, clothes, a third, the warmth of the hearth. An item from Tajikistan stated that when the children asked their mother to give them more pancakes for supper, the mother answered, "If I give you more, what shall we leave for the adjina? She will come at night, and finding nothing may become angry."

There are stories in Tajikistan that when the cry of an infant is suddenly heard from a barn, it means that a demon has given birth. People give food to her, "she eats, takes the baby, and goes away."

In Georgia, the ancient clan of Naraani was said to have befriended a dev. They "fed him well," leaving food warm in the ashes of the hearth. When the family went to sleep, he would come and have his fill. If food is not offered, the demons would steal it—all kinds of it—especially vegetables and fruits from gardens and orchards.

As a rule, demons are seen naked, but there are many exceptions, and clothing is the next item of interest motivating contact with humans. It is advised, when encountering a goblin in the woods, to offer it bread or a piece of clothing, even a torn-off sleeve if nothing else is available. On record are Ukrainian and Bielorussian songs telling how rusalkas beg human girls to give them shirts, no matter how old or tattered. No wonder, demons usually sport threadbare garments, often worn the wrong side out. As a result, when Russians saw a man in a shirt worn inside out, they used to say: "Look, he is [dressed] like a leshy!"

The leshy were said to approach campfires built by lumberjacks or hunters in order to warm themselves in cold weather, and it is said that they "turned away their muzzles," apparently because of the bright light. They also took care that flying sparks did not touch their hair.

Seeking warmth, they also entered peasant bath huts or barns for crops stocked there. It is reported that a leshy, festooned with icicles, entered a barn and put out a fire with melting ice. In contrast, in the summer they would come up to a campfire not for warmth, but to put it out.

Folklore is insistent that demons love human children. Hiding from adults, they often come in view of children and even play with them when adults are not around. They are also said to calm down crying babies and, inevitably, as a result of such fondness occasionally take human children with them. In Bielorussia, a wood goblin was "charged" with stealing a cradle with a baby and hiding it in a birch tree. In the Novgorod province, a boy of 13 was kidnapped by a wood goblin. Four years later the boy returned naked and unable to speak.

The demon's voice is usually described as "vociferous," and their sound mimicking ability is often mentioned. In Russia, for example, the leshy is said to be able to imitate the voices of human males, females, and babies; he can neigh like a horse, squeal like a pig, bark like a dog, meow like a cat, and crow or cluck like a cock and hen.

Demons are mentioned in proverbs and sayings, which people still use commonly today. Every proverb has two meanings: one direct and literal, concerning real life; the other indirect and figurative, alluding to people's behavior. Thus, when we say "A bird in the hand is worth two in the bush" or "One shouldn't look a gift horse in the mouth," we use literal, real-life meanings in a figurative way. So what is the real life meaning of the numerous proverbs and sayings referring to the devil and other demons?

The Russians say "The devil is not so ugly (or fearsome) as he is painted." The English say, "The devil is not so black as he is painted" or "To paint the devil blacker than he is." The Russians also say, "The devil is swarthy from birth, not from the sun," and they say, "Brown devil, gray devil, still a devil." This means that the creators of these proverbs were familiar with the look of the devil.

The Russian equivalent of the English, "Still waters run deep," is, "Devils dwell in a quiet slough (pool)." The famous 19th Century lexicographer, Vladimir Dahl, offers other proverbs and sayings reflecting the devil's aquatic preferences: "To be led to the devil, like the devil to the marsh;" "Given a marsh, given the devils;" "When devils dive nothing but bubbles arise;" "A job (a work assignment) is not a devil, won't disappear into the water;" and finally, "Worms in the earth, devils in the water, crooks in the court, where can a man go?"

Some more sayings from Vladimir Dahl's *Dictionary of the Russian Language*: "You are as big as the devil (or leshy) but still small in the mind;" "You are clever and strong but can't beat the leshy;" "Leshy is mute but vociferous;" "To roar like a leshy;"

"Infected with the devil's fleas and lice;" and finally, "The devil brushed himself and lost his brush."

An Arab proverb states, "Azrata min ghoul" (stinking like a ghoul); also quite a familiar sign. A synonym for "demon" in Russian is "unclean spirit." Demons collectively are referred to as "nechistaya sila" (unclean power). When the Kabardians say, "to catch the almasty by head hair," they mean to pull a thing off. The advice and wish, "Go to the devil!" and, "The devil take you (him, her)" seem to be international. When a needed person appears at last after a long wait, the Russians say, "Where has the devil been carrying you?" Enlightened by the Albert Ostman case[1], the hominologist knows that the latter saying is a reflection of real life as well.

There are many examples of demon killings in folklore. According to one item from Siberia, a reduction in wood goblin numbers there was due to the appearance of hunting guns. Some tales relate that hunters, having killed a demon, cut off parts of its body (sometimes the head) as souvenirs and valuable trophies. Obviously, encounters with human beings wielding firearms boded no good for "mythical beings" and this is a reason for their legendary seclusion.

There are also plenty of beliefs that demon killers suffer inevitable retribution for the deed. Chuvash folklore intimates that in a village where "upate" (literally half-man) were killed, human population no longer increased. Tatars had similar beliefs, and when they saw a little poor village, they used to say, "Shurale kargagan" (condemned by shurale—the latter word meaning wood goblin). An example from Azerbaijan mentions a hunter who fired point-blank at a "biaban-ghouli" who fell to the ground, then stood up and ran away, leaving behind a bloody trail. The hunter sold his gun and never hunted again. Asked why, he answered, "After that all my children died." A parallel First Nation tale was published in 1929 in Canada by J.W. Burns and reprinted by John Green in *The Sasquatch File* (1973: 11).

1. In 1957, Canadian Albert Ostman testified before a magistrate of being carried off by a sasquatch some thirty-three years previously. He claimed to have been held with a family of four sasquatch (man, woman, and two children) for six days before he managed to escape and return to civilization. His full account can be found in John Green's, *Sasquatch: The Apes Among Us*, 1978, pp. 97-110.

Cases of demons imprisoned by humans are also numerous in folklore. A creature, especially young, could get entangled as already mentioned, in a fishing net. To catch migratory birds, the Russians used to hang a huge net on the trees of a forest vista. It happened sometimes that instead of wild ducks and geese, the hunters found a devil in the net. The technical term for this kind of net is "pereves." So there appeared a proverb, "popalsya kak bes v pereves," (caught like a devil in a net).

From Tatar folklore we learn that the inhabitants of a village tired from the tricks of a shurale (wood goblin/demon) who troubled their herd of horses every night. They spread tar on the back of their best horse and by this ploy caught a she-demon who had tried to ride that horse. But the best and most ancient method of catching demons was by intoxicating them with alcohol. In ancient Greece it was used by King Midas to catch Silenus, a forest spirit. In a Temple of Silenus "drunkenness is represented in giving him wine in a cup" (Pausanias VI, 24). In Italy, alcohol was used by King Numa Pompilius to catch a faun (forest god or goddess).

Being so rare and impressive, these events were recorded by legend. The only modification in the method in Russia is that wine is replaced by vodka. A tale from Abkhazia has it that a wood goblin who meddled with hunters' traps was caught only after imbibing a bucketful of vodka.

Of special importance among the sources of information is the Bible. The beings of our interest are mentioned, for example, by Isaiah in his prophecy against Babylon. The prophet says that Babylon, the glory of kingdoms, shall be destroyed, turned into a waste land, and "wild animals of the desert" shall come to live there. Along with such denizens of the desert as ostriches, jackals, and hyenas, the Bible in Russian mentions the leshy (wood goblin).

How came wood goblins to be in the desert? The discrepancy intrigued me and demanded an answer. In search of it, I discovered that the earliest edition of the Bible in Russia specifies the "devil" instead of "wood goblin" in those verses of Isaiah. I then looked up the Authorized Version of Isaiah in English and discovered "satyrs" in the corresponding verses. So I opened the *Encyclopedia Britannica* (1965, Vol. 20: 11), and read in part the following:

Satyrs, in Greek mythology, spirits half-man, half-beast. (...) In Italy often identified with the fauni. In the Authorized Version of Isaiah 13:21; 34:14, the word 'satyr' is used to render the Hebrew se'irim (hairy ones)—a kind of demon or supernatural being, known to Hebrew folklore as inhabiting waste places. They correspond to the "azabb al-'aka-ba" (shaggy demon of the mountain-pass) of old Arab superstition."

So what were the "hairy ones," alias the "shaggy demon of the mountain-pass," alias wood goblins, alias satyrs, alias devils, doing on the ruins of Babylon? Various translations of the Bible into English answer as follows: they "will leap about," they "will dance," they "shall call to each other," and finally, they "shall cry out one to another." Well, I thought, Isaiah could well be considered a forerunner in the field of hominology. After all it was not he who called the "hairy ones" by such names as "goblins," "satyrs," and "devils." He used the term derived from the creatures' biological characteristic, i.e., their hairiness.

I then turned to the New International Version of the Holy Bible, and found an alternate rendering. Here "wild goats" were mentioned instead of "satyrs"— "And there wild goats will leap about..." Also, "and wild goats will bleat to each other." What a leap from the original intent of the reference!

The erotic aspect of hominology is reflected most prominently both in ancient literature and world folklore on the subject. According to legend, the Babylonian King Gilgamesh habituated and befriended the half-man, half-beast Enkidu with the help of the priestess of the goddess of love, Ishtar. Enkidu is

Figure 9. Assyro-Babylonian demon Humbaba, lord of the cedar forests in the mountains, who did not allow people to cut them. (Photo: Public domain)

said to have been shaggy with "hair that sprouted like grain." He ate with the gazelles and drank with the wild beasts at their waterholes. He protected wild animals from hunters, so a hunter went to King Gilgamesh with a request for help. The king recommended that the hunter take a priestess of Ishtar with him to the waterhole and instruct her to disrobe, thus enticing Enkidu away from his animal friends. The ruse succeeded and the wild man enjoyed the woman's favors for a week, being gradually persuaded to eat bread and drink wine with the shepherds. He became their friend and helped them by driving lions away from their flocks. Subsequently, Enkidu found himself in the palace of Gilgamesh and became the king's best friend and aid in hunting. He also helped Gilgamesh in fighting the monstrous demon Humbaba, actually another wild man, in the forested mountains of Lebanon (Reder, 1965). (Fig. 9)

Lustfulness is a distinguishing trait of satyrs in ancient Greece. Ancient historian Diodorus Siculus (c. BC 90–21.) wrote about satyrs: "this **animal** (emphasis added – D.B.) shamelessly seeks crossbreeding"[with humans] (Diodorus, 1774). Pausanias, in already quoted *Description of Greece,* citing Euphemus regarding the danger encountered by mariners on the Isles of the Satyrs, inhabited by "wildmen," says that the satyrs "ran down to the ship, and without uttering a syllable attempted to get at the women in the ship. At last the sailors, in fear, cast out a barbarian woman on the island, and the Satyrs outraged her most grossly" (Pausanias, 1913: 33). As a result, European languages have acquired such ancient medical terms as *satyriasis* and *nymphomania*.

Among the commandments that Moses gave to Israel was: "And they shall no more sacrifice their victims to devils, with whom they have committed fornication. It shall be an ordinance forever to them and their posterity" (Leviticus 17:7, The Holy Bible, Douay Version, reproduced from the first edition of The Old Testament, printed at Douay in 1609).

Another translation in *The Holy Bible,* London, 1850 reads: "And they shall no more offer their sacrifices unto devils, after whom they have gone a whoring. This shall be a statute forever unto them throughout their generations." A third version, published in The New English Bible, Oxford, 1970 reads: "They shall no longer sacrifice their slaughtered beasts to the demons whom they wantonly follow."

Let us note that, according to the Hebrew text, Moses did not use the words "devils" or "demons" in this commandment by the Lord. Again the term *se'irim* (hairy ones) was used, which presented a sticking point for the translators. "Hairy ones," and moreover sacrifices to and fornication with them, called for an explanation; "wild goats" would not fit in this case. So were "devils" and "demons" found to be preferable terms for who do not know that devils and demons are seducers and corruptors of humankind?

Christianity also condemned "pagan gods" for lustfulness. "St. Augustine in his City of God brackets fauni and sylvans together as incubi, and then proceeds to explain that both ''desire women and act carnally with them" (Bernheimer, 1962: 97; Civitas Dei, 15, 23).

In 1484, "Pope Innocent VIII issued a [papal] bull against witches. It has been estimated that during the next three centuries 300,000 to 2,000,000 persons were executed as witches" (*The Encyclopedia Americana*, 1973, Vol. 29, Witchcraft). A standard charge by the Inquisition for its victims was intercourse, including sexual, with a demon. In this connection of special interest is the work by Italian theologian, Luigi Maria Sinistrari (1622–1701), jurisconsult of the Inquisition's High Tribunal in Rome. Sinistrari argued that it was necessary for the Inquisition to distinguish between culprits, who associated with real demons, and people who fell victim to certain man-like animals, mistaken for demons. Accordingly, Sinistrari's work has a long and instructive title:

> *On demonism and the animals, incubi and succubi, where it is proved that there are reasonable creatures on Earth, apart from man, which have like man a body and a soul, which like man are born and die, which are redeemed by our Savior Jesus Christ and capable of salvation and damnation.* (Sinistrari, 1875; my translation from the French).

Sinistrari mentioned popular names of these animals, such as folletto in Italy, follet and lutin in France, and duende in Spain (all translated as "goblin"). His main argument for why these beings are animals, not evil spirits, is this—they are immune to exorcism. It happens, he wrote, that they "meet exorcism with a grin," or "even beat up exorcists and tear up sacred clothes." Hence, it is clear they "are not evil spirits or angels, nor are they human beings, even though they are endowed with reason."

Further biological traits of these animals, pointed out by Sinistrari, are the following:
—They seek sexual intercourse with humans;
—From such intercourse children are born who, when grown up, become very tall, strong and daring;
—These animals' vocalization resembles whistling;
—These animals are attracted by horses and like to plait their manes (this has been reported elsewhere in Eurasia and the Americas – D.B.);
—These animals throw stones and pile them up;
—It is very difficult to see these animals, being seen either by chance or of their own volition;
—They are capable of feeling and suffering, but being very swift and nimble in avoiding danger, it is surprising that they get killed or injured at all. This can happen when they are asleep or in some other inadvertent way.

Thus we see that the biological nature of the creatures, regarded as evil spirits by some and as figments of the imagination by others, was apparent to a theologian consulting the Inquisition. We do not know if Sinistrari's distinctions saved any people from death during the witch-hunt in Catholic countries, but in Lutheran Sweden death sentences were passed even in the 18th century for sex with female trolls, called *skogsra*. I have this information from the late Norwegian hominologist Erik Knatterud. He has uncovered references to court documents of the 17th and 18th centuries regarding these cases (personal communications of June 23 and July 2, 2003).

Incidentally, in 1990, I was in contact with a university student in Sweden, Niclas Burenhult, who was studying cultural anthropology, and wrote me that, "In 1555 Olaus Magnus published his *Historia de gentibus septentrionalibus*. (...) Olaus Magnus was the last Catholic archbishop of Sweden. This work is said to be a unique insight into medieval Scandinavia. The author travelled throughout Sweden and described the geography, animal life, history, traditions, etc., of the country. In a sense he was a sort of early Linne" (Linnaeus – D.B.). Magnus

(1490-1558) happened to touch on the subject of trolls, and Burenhult translated for me from Swedish the following extract:

> It is an established fact that the inhabitants of the north experience great services and assistance from the trolls. This is most often occurring in stables and mines. In the latter the trolls usually dig out, hollow and cleave blocks of stone and load them in barrels. (...) Other trolls are very harmful, like the one who ran so wild in the mine at Anneberg a few years ago that it slayed twelve miners. (Niclas Burenhult, personal communication of October 9, 1990).

As for Russia in past centuries, I am not aware of court actions against people interacting with homins, but I know that the Orthodox Church regarded any such relationship as a great sin. This attitude is reflected in Bielorussian folklore by an incantation, an enchantment, which is a kind of instruction to a young peasant telling him what to do if he is accosted by a rusalka, a female homin. It is pointed out that the man should not look at her but at the ground, and say the following:

> Water dweller, wood denizen, wild, unruly and whimsical girl! Go away, get away, don't show up at my homestead! (...) I kissed the golden cross and abide by the Christian faith, so can't mix with you. Go to the pine forest, to the forest lord. He has prepared a bed of moss and grass and is waiting for you. You are to sleep with him, not with a Christian like me. Amen. (Shein, 1893).

Sexual relations of humans with demons is a topic present in many works on folklore and natural history that I read and referred to in my book. The 12th century Persian scholar Nizami al-'Arudi, mentioned earlier, wrote that "the Nasnas (...) is very curious about man. (...) And if it sees a lonely man it abducts him and is said **to be able to conceive by him**" (my emphasis – D.B.). (Nizami, 1963).

One such success story in crossbreeding is reported by Kazakh folklore, telling of a horse herdsman who encountered a female almasty (wildwoman) in the steppe and thought, "Be it a shaitan or a human, it doesn't matter." He cohabited with her and "they had three children born to them."

If we give credence in this respect to folklore, then hominology is faced with the question: What is the genetic status of "demons," i.e., homins, in relation to *Homo sapiens*?

"Good" species are not supposed to produce fertile crossbreeds. Still, division into species and subspecies of closely related organisms is often a matter of speculation and consensus. Primatologists are aware of fertile hybrids of different monkey species. Another case in point is the example of wolves and coyotes, considered to be different species. Yet they carry the same number of chromosomes and there exists no genetic barrier to their interbreeding. If not for behavioral differences, which keep them separate, one species would have long ago absorbed the other.

The homin-human situation appears to be similar; the barrier to crossbreeding is likely behavioral, not genetic. On this basis it could be overcome in principle and in practice, but the process has been censored and censured.

The history of man's relations with homins is full of ambivalence. The wild hairy bipeds were believed at one time or another, or simultaneously, to be gods, demi-gods, devils, half-men and wildmen. Accordingly, views on their gifts and abilities have been varied and often contradictory. One exception however is the unanimity of opinion regarding their physical endowment. All popular demons of both sexes are far more athletic than humans. Many folk tales relate of physical competitions between man and demon, and every time man would resort to ruse and trickery to "win" the round.

On record is Pliny the Elder's phrase in *The Natural History*: "the Satyrs have nothing of ordinary humanity about them except human shape." This hominologist tends to both agree and disagree with the ancient scholar. The beings in question seem very different from ordinary humanity, and at the same time they are like human beings not only in shape but in many other respects as well.

The ancients believed satyrs to be gods and demi-gods, which did not prevent Hesiod from saying that these "brothers of mountain nymphs (were) an idle and worthless race" (Strabon, 1964). If this means that satyrs and their ilk do not earn a living by labor, it is correct. For all we know today, they lead an animal way of life.

We also know today that some animals make and use tools that help them obtain nourishment. How about demons in this respect? There is mention of clubs in the hands of wood goblins, but no mention of stone tools; more often use of stones as projectiles. There are also references to tools taken from man. Rusalkas, for example, were seen with a pestle in hand; they were often described combing their hair with combs, apparently taken from peasant bath huts which they visited.

A peasant once observed a rusalka standing in the water and looking into it as if into a mirror, smartening herself up. This indicates a level of self-awareness shared only by humans. Other accounts suggest that rusalkas used to erase their footprints on a sandy river bank. Folklore avers that they make wreaths of flowers, sedge, and tree branches, and put them on their heads. Let us also note that satyrs, nymphs, fauni, etc., are often depicted adorned with wreaths.

Pan, the god of flocks and shepherds, when tired of striking panic into man, would start playing a flute. There are also pictures of satyrs on Greek vases doing the same. Pan is even credited with inventing the shepherd's flute, the syrinx. Satyrs, nymphs, *Oriental peris*, and Russian rusalkas love dancing and merrymaking, which is credible enough, but I always doubted that demons not only dance but also play music and invented a musical instrument. So I wondered why the Greeks credited them with such gifts. Then I happened to read Dr. Henner Fahrenbach's report on sasquatch imitating "even short phrases on a flute." Indeed, the sasquatch has traditionally been associated with a whistling call. This prompted me to think that when a Greek shepherd played a flute, Pan and company, well hidden in the woods, simply imitated the sounds, and hence the origin of the legend.

Demons can wear clothes, given by humans or stolen from them. The clothes are usually old, tattered, and worn inside-out. There is mention of wood goblins tearing off bast from trees and trying to make bast shoes, maybe in imitation of similar work by peasants. One item tells of a rusalka that made a cradle for her baby out of birch tree bark. In this connection let us recall Albert Ostman's words about sasquatch: "... they had some kind of blankets woven of narrow strips of cedar bark, packed with dry moss. They looked very practical and warm—with no need of washing." (Green, *Sasquatch: The Apes Among Us*, 1978, p. 105)

There is mention of various activities of demons helping humans—in hunting, fishing, pasturing, as well as in household work. Such activities are viewed very positively in folklore, with only a few exceptions. For example, regarding the Georgian dev mentioned earlier, it is said that when people were making hay on a hill, during the night the dev carried all the haystacks to the hilltop, while hay was needed in the valley below. "The people thought to themselves: 'Why wouldn't he carry the stacks down instead of uphill?' The next night the dev brought all the hay down."

The work of household she-demons is highly praised, but is noted that they cannot bake bread because they burn their hands. In regard to fire, it is clear that demons are not afraid of it. They approach campfires and hearths to warm themselves and they are able to put out fire, but are never said to be able to make it.

Demons can laugh; in sorrow their women and children would weep. They can sing, whistle, and imitate cries of various animals and voices of people (males, females, and babies). As for the crucial question of speech, the answer in folklore is usually negative. The Jewish Talmud recommends a method for detecting a demon in the dark. If you happen to run into someone in the dark, the Talmud recommends saying "Shalom!" (Hello). If the greeting is not returned, chances are you are facing a demon (*The Universal Jewish Encyclopedia*, "Demons"). The same device is mentioned in Georgian folklore, using Georgian "Gamarjoba!" instead of "Shalom."

Folklore mentions demons resorting to gestures and fingers when communicating with humans. Vladimir Dahl writes that demons "sing without words," that their mumbling heard from a distance can be taken for speech, and peasants would interpret it in a jocular way (as if meaning "walked, found, lost" or "worse off every year"), but when coming face-to-face with a demon it would become clear that he is speechless.

But if homins have not crossed the "Rubicon of mind" associated with the faculty of speech, there is little doubt they have come close to it, and some may even have stepped into it. This follows from evidence in different habitats (central Russia, the Caucasus, Tajikistan, China, North America) of their sound-imitating ability, which is *conditio sine qua non* in the origin of speech. And it is noteworthy that if not all, then at least some homins have been reported to be capable of utterances resembling words. No ape is capable of such mimicry, being capable of mastering only the simplest of monosyllabic utterances such as "ma."

Concluding Remarks: Folklore is a rich source of information for the discerning hominologist and, at the same time, an obstacle that has to be overcome on the way to the truth. My book *Wood Goblin Dubbed Monkey* serves this double aim. It concludes with the question: "Will goblins help the world of science to open its eyes on what was clear to Boris Porshnev over twenty years ago?"

Folklore suggests that human-wild man relationships throughout the millennia has been one of a love-hate kind. The lords of nature have been deified and condemned, offered sacrifices and hunted as valuable game for food and medicinal purposes. We also learn that various specimens have been captured, tamed and exploited as warriors, hunters, and unskilled laborers. Why then have they not been turned into slaves, or a kind of most sophisticated domestic animal? Wolves have always been man's enemies, but transformed into dogs have become man's best friends. Why hasn't this happened with pre-human homins?

I think the answer is evident: genetically they are so close to humans that they tend to interbreed with our kind. But, unlike human slaves, they are unable to understand and obey human rules and customs of sexual behavior or respect a ban on interbreeding with humans. This may explain the homins' special role and place in human history and culture. Perhaps for this reason information about our wild hairy cousins has been concealed and kept secret throughout ages, and also why it has been so greatly mythologized, having reached us abundantly by way of myth and folklore and much less by way of natural history and science.

Cryptids are usually hidden in forests, mountains, lakes and oceans. The object of this research is likely hidden in natural forests and mountains, but above all it lurks hidden in the "forests of the mind." If not for these forests, the problem might have been resolved long ago. The task of hominology is to take the creature out of these dark forests and into the light of objectivity.

One final relevant question: How to correlate relict hominoids with the fossil record of paleoanthropology? According to Dr. Grover Krantz (1980), Neanderthals had more traits in common with *Homo erectus* than with *Homo sapiens*, so that they "could all be classed with *erectus*." He also wrote: "*Homo erectus* existed for over a million years with relatively little change—a kind of evolutionary plateau—and then was transformed rather quickly into *Homo sapiens*" (Krantz, 1980).

Could it be that today's wild bipedal homins are relicts of that evolutionary "standstill" which lasted long enough for them to penetrate and settle the Old World before the advent of *Homo sapiens*? Adapting to local environments, these pre-sapiens must have more or less departed in their physique from the fossil erectus-grade forms presently known to science. I therefore propose that homins reported in central Eurasia are relicts of the *Homo erectus*-Neanderthal stage of evolution.

Lastly, some concluding thoughts from my address at the International Bigfoot Symposium in Willow Creek, California, in September 2003. I think that one of the great scientific results of the 20th century was the discovery of relict hominoids, popularly known as abominable snowmen, yeti, yeren, almas, almasty, bigfoot, sasquatch, etc. Actually, it was a re-discovery by hominologists of what had been known to western naturalists from antiquity to the middle of the 18th century, when wild hairy bipedal primates were classified by Linnaeus as *Homo troglodytes*.

As for Eastern scholars and rural populations in many parts of the world, they have always been aware of wild hairy bipeds, known under diverse popular names. Thus, on the agenda is not their discovery, but general recognition of their re-discovery in the last century. Such recognition is expected to make a tremendous impact.

Postscriptum: In the spring of 2002, I was contacted by Mary Green and Janice Carter Coy, Tennessee, USA, who asked permission to use material from my book, *In the Footsteps of the Russian Snowman*, in the book they were writing. Permission given, a lively exchange of information followed between the ladies and I which overturned some of my views on hominology. The reason can be seen from the very title of the book that Mary and Janice published in December 2002, *50 Years With Bigfoot: Tennessee Chronicles of Co-Existence*.

Written by two lay people, the volume, devoid of scientific methodology and analysis, is none-the-less absolutely sensational because of its subject matter. A bigfoot family had lived and co-existed for many decades with the Carter family on their wooded farm on the edge of the Smoky Mountains. From her childhood on, Janice had observed the anatomy and behavior of this family of giant North American homins and offered in the book her detailed observations.

Most American hominologists have rejected the book because it contradicts their views and convictions, so the story is still in limbo. In contrast, I embraced and accepted it from the beginning because Janice's account was in perfect agreement with what I learned from local people during my expeditions in Russia and in particular from folklore described above. Both from folklore and the locals I learned of cases of co-existence and even friendship and cooperation between homins and humans in the manner described by Janice.

So why were some of my views overturned? Because previously I had believed that homins are devoid of the power of speech and language; therefore I classed them beyond the level of human beings. Janice firmly insisted on her bigfoot's peculiar ability to speak and even learn English. This creates a "revolution" in hominology, indicating that we are dealing not with "beasts" or "superanimals," as I referred to them, but with a kind of human being whose psyche and mental abilities we are still far from knowing and understanding.

CHAPTER 2

Learning from Folklore

(Originally published and posted in a series to Bigfoot Encounters website, December 2009 through February 2011; currently on the Sasquatch Canada website.)

That devils and wood goblins could be the names of real beings, I heard for the first time in 1964 from Professor Boris Porshnev. In the summer of that very year I joined an expedition to the Caucasus, and that was the time of legend come to life for me; a very memorable and amazing event indeed. One thing is to read it in a book, and quite another to hear the local people use matter-of-factly such names for the hairy bipeds they encounter.

Back in Moscow, I went to the best public library and delved into books of folklore and demonology. Some time later this resulted in a manuscript, "In Defense of Devilry," approved by Porshnev and a friend of his, an outstanding ethnographer; but (as previously noted), nobody wanted to publish my work.

With political changes in the country—the perestroika (i.e., back to capitalism)—I expanded the manuscript and changed the title (the church was getting more say in politics) to *Wood Goblin Dubbed Monkey: A Comparative Study in Demonology*. As epigraphs I used the following remark by a satirist, "Many things are incomprehensible to us not because our concepts are weak, but because these things are not covered by our concepts," and Thomas H. Huxley's thoughts in his book, *Evidence as to Man's Place in Nature*, 1863: "Ancient traditions, when tested by the severe processes of modern investigation, commonly enough fade away into mere dreams: but it is singular how often the dream turns out to have been a half-waking one, presaging a reality."

I offered the work to many publishers without success, but finally it was published in 1991, the year of the Soviet Union's disintegration. The economic situation was catastrophic, so nobody cared for wood goblins and the like. I sent copies to several scientists and received laudatory responses, but to this day not a single review of the book has appeared in print.

My approach to folklore as a source of information in our research was explained in the article, "A Note on Folklore in Hominology," published in *Cryptozoology*, Vol.1, 1982. It's worthwhile to quote it here. The relevant term I used at the time was "hominoid" (not hominid) in its literal sense of "man-like being," not in precise terms of taxonomy. Those who took wild hairy bipeds for biological beings I called "realists;" those who regarded them as mythological images I dubbed "folklorists." So here goes:

> The relationship between "realists" and "folklorists" in hominology has not been easy or productive, and this has induced me to re-examine its background and to try to lay down some basic rules. There are philosophers who insist that "reality" exists only in the mind of the beholder. I know of no logical argument to counter this assumption, which can be regarded as an extreme case of "folklorism." Presumably, such a philosopher, if kidnapped by a sasquatch, like Albert Ostman was, would be consoled by the thought that the drama is only taking place in his head.
>
> On the other hand, we know that an archaeologist, Heinrich Schliemann, who, proceeding from the ornate imagery of the ancient Greeks, confronted the world with the reality of Troy. Schliemann was a realist, and there can be little doubt that if he and other archaeologists had asked and followed the advice of "folklorists" on the reality of Troy, its precious relics would still be lying underground.
>
> This example shows that there can be totally different entities bearing the same name, and our failure to recognize and differentiate such entities leads to a lot of confusion and useless arguments. The name Troy applies, on the one hand, to a figment of an ancient poet's imagination, studied by specialists in literature and mythology, and, on the other hand, to a real historical city, whose study is the business of archaeologists and historians.
>
> Of course, the two entities are interconnected in some way; one was the cause of the other, and both can have some overlapping characteristics, but on the whole, their natures are so different that it would be most unwise to judge the one, say the historical city of Troy, by our knowledge of the other, the mythological Troy.
>
> I believe the same analogy applies in hominology, the

term we applied to the study of sasquatch-like creatures. There are real hominoids (that is, creatures of biology—we know this from several categories of evidence combined), and there are imaginary ones (those of mythology). Our opponents say that one kind is quite enough (those of mythology), which dispenses with the necessity for real ones. But I say nay—the existence of mythological hominoids is a necessary, though not sufficient, condition of the existence of real hominoids. The argument was set forth by us in 1976 as follows:

Folklore and mythology in general are an important source of information for science. But hominologists look for myths about these creatures not only to find a real basis for the myths and to supplement their knowledge of the problem. They also need the myths as such, for they are yet another "litmus test" confirming the historical reality of hominoids. If, in the course of history, people had encounters with "troglodytes," then these most impressive beings could not have escaped the attention of the creators of myths and legends. Of course, the reality of relic hominoids cannot be supported by recourse to folklore alone, but neither can it be refuted by such references, as our opponents have attempted to do. Is the abundant folklore, say, about the wolf or the bear not a consequence of the existence of these animals and man's knowledge of them? Therefore, we say that, if relic hominoids were not reflected in folklore and mythology, then their reality could be called into question. Fortunately, this channel of information is so wide and deep that much work can be done in the sphere; it is necessary to re-examine and re-think a good many anthropomorphic images playing important roles in folklore and demonology.

The last sentence above seems to find support in the words of Wayne Suttles, a cultural anthropologist at Portland State University: "If there is a real animal, shouldn't there be better descriptions in the ethnographic literature?"(...) Not necessarily. Anthropologists do not consciously suppress information, but they sometimes **do not know what to do with it.** There are ethnographies of peoples whom I know to **have traditions of sasquatch-like beings that make no mention of such traditions;** I suspect that these omissions occur not because the writers had never heard of the traditions but because they **did not know how to categorize them.**" (Suttles 1972).

Why is it difficult for ethnographers to categorize such material? Probably because they have no idea what is real and what is imaginary in it. And the fact that the informants do not know either cannot be of much help to the scientist, who should always attempt to draw a line between fact and fiction.

Hence, ideally, "realists" and "folklorists" in hominology should sit down together and, without violating each other's territory, sort out the mountain of folklore on hominoids. When Suttles says that "a large non-human primate would not really steal women" (Suttles 1972), I am afraid he trespasses on the turf of other kinds of experts. When a nineteenth-century Russian ethnographer said that the large breasts of a female wood-goblin (forest woman) had been made up by ignorant peasants to symbolize heavy precipitation, he simply ascribed his own ignorance and fantasy to his informants. What about the image of a "tree-striker" that has the habit of "knocking down dead trees"(Suttles 1972)? Well, if it's a hominoid's way of feeding on larvae, the image has a basis in reality. (My emphasis).

Wayne Suttles, authored the article "On the Cultural Track of the Sasquatch" in the journal *Northwest Anthropological Research Notes,* 1972, reprinted in *The Scientist Looks at the Sasquatch,* 1977, a collection of articles, edited by Roderick Sprague and Grover Krantz. The volume had a second edition in 1979, with added articles, including "The Improbable Primate and Modern Myth" by Richard Beeson, University of Idaho. He wrote in it:

Even more incredible are the majority of reports of sasquatch females which time after time describe these animals as having large, hairy, pendulous breasts (Green 1970:77; 1973:50). Let us look again at the female sasquatch. It is reported to be both very hairy and to possess large pendulous breasts. One is about as likely to find that combination in the order of primates as a fish on a bicycle. Among primates, large pendulous breasts are indicative of a level of sexual sophistication that can only occur in very intelligent, symbol-using animals. Man is one such animal; the sasquatch is not. (pp. 175, 176).

This reminds me of the "blast" by Dr. William Montagna, director of the Regional Primate Research Center at Beaverton, Oregon, aimed at the Patterson/Gimlin documentary footage, which he called "this few-second-long bit of foolishness;" "The crowning irony was Patterson's touch of glamor; making his monster into a female with large pendulous breasts. If Patterson had done his homework, he would have known that regardless of how hirsute an animal is, its mammary glands are always covered with such short hairs as to appear naked" (*Primate News,* Vol.14, No.8, September 1976).

As to Richard Beeson, his conclusion was as follows:

To summarize and conclude: we have examined the existing literature containing several hundred first-hand reports of the Sasquatch. These reports present the physical and behavioral profile of an animal whose essential traits are for the most part highly improbable and, in respect to some, entirely impossible. (...) What the Sasquatch represents, I believe, is **a modern form of myth** and we are privileged to be able to see it in the making. (*The Scientist Looks at the Sasquatch II,* The University Press of Idaho, 1979, pp. 192, 193, my emphasis – D.B.).

I imagine how in the future the above will entertain and instruct students of science, having become classic examples of ill-considered judgment in science. Today, 30 years on, most cultural anthropologists remain as skeptical as Richard Beeson was, but at least three, known to me in America, have become full-fledged realists in the sense I indicated. One of them is Kathy Moskowitz Strain. In May 2008 she gifted me her book, *Giants, Cannibals & Monsters: Bigfoot in Native Culture* (Hancock House, 2008). I greatly enjoyed reading it in the summer, but did not have time to comment until much later.

The book is marvelous not only for its stories but also for its numerous photographs showing the various Native people of North America. In Acknowledgements Kathy writes that "Christopher Murphy worked very hard on the layout of the book." Chris, in turn, wrote me, in part, about his work on this volume:

> As I read it, I visualized the stories being told by Natives around a campfire with wide-eyed children transfixed with the story-teller. It was then that I realized that the book must contain images of the different people in their regular walks of life, all placed with their stories. In this way the reader would get a better appreciation of the stories, and at the same time realize just how diverse the Native people are in North America. (...) We are fortunate that there were early photographers who liked to take photos of Natives—who by the way are highly photogenic. I would say generally that the photos of Native people are 80-100 years old or older. I don't think any Native seen is now alive, even the wonderful little kids and young people.

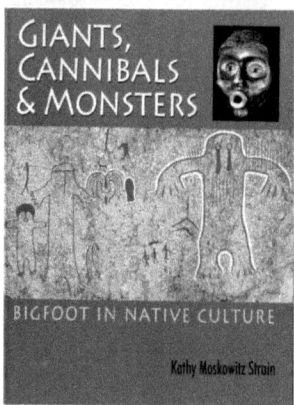

Kathy Moskowitz Strain and her remarkable book. (Photo of Kathy: C. Murphy)

So Kathy Moskowitz Strain has done for North America in the field of Native folklore what I've done for the former Soviet Union. The difference though is in the presentation of material. I grouped different ethnic tales according to their similar or identical description of one or another trait of "wildmen" and "demons" (their appearance, food, behavior, etc.). Kathy presents Native tales, one after another, as told by members of this or that Native people, so I have followed her order of presentation in the analysis that I now present.

Notes on my analysis: Phrases and words of particular interest in the quoted material are underlined. My comments or commentaries are shown below the quoted material. Where a page number is shown on the first line, that is the page number in Kathy's book when applicable. Original sources for all material used from Kathy's book are shown in full (book, author, date, and so forth) after the entries. If no source is shown, then the source is Kathy Moskowitz Strain.

In many cases, I need to refer several times to my books or certain books written by others in my comments or commentaries. To avoid wordy repetition, I have shown these books with the following abbreviations, followed by page numbers:

FRS = *In the Footsteps of the Russian Snowman*, Dmitri Bayanov, 1996, Crypto-Logos, Moscow, Russia.
BRV = *Bigfoot Research: The Russian Vision*, Dmitri Bayanov, 2011, Hancock House Publishers, Surrey, BC.
SLS = *The Scientist Looks at the Sasquatch*, Edited by Roderick Sprague and Grover Krantz, 1977, University Press of Idaho.
THP = *The Hoopa Project*, David Paulides, 2008, Hancock House Publishers, Surrey, BC.

The Shasta (page 51)

The *Tah-tah-kle'-ah* (Owl-Woman Monster). Before the tribes lived peaceably in this country, <u>before the last creation</u>, there were certain people <u>who ate Indians whenever they could get them. They preferred and hunted children, as better eating</u>. These people, the *Tah-tah kle'-ah*, <u>were taller and larger than the common human. They ate every bad thing known such as frogs, lizards, snakes, and other things that Indians do not eat. They talked the Indian language, and in that way might fool the Indians</u>. (...) But <u>at the last creation</u> they came up only in California. Two were seen there. <u>They were women, tall big women, who lived in a cave</u>.

Original Source: *Ghost Voices: Yakima Indian Myths, Legends, Humor, and Hunting Stories*, by Donald M. Hines, 1992. Issaquah: Great Eagle Publishing, Inc., pp. 63-64.

Comments: I wonder what "the last creation" means? Mention of homin cannibalism is ever present in North American Native folklore, much more so, to my knowledge, than in Eurasia. Wonder why? Note that cannibalism is found to have been practiced by both Neanderthals and *Homo sapiens*; in the case of the latter in some indigenous cultures even today. Does "Owl" mean *Tah-tah-kle'-ah* were nocturnal? Their body size, food, and dwelling are all realistic. That they "talked the Indian language" is a stumbling block and pressing question. Did they really talk it or only pretend in order to "fool the Indians?" Or both? Today, a sasquatch, uttering Indian words, is by no means "a fish on a bicycle" for me. And what an amount of valuable information in such a short passage!

The Wintu (pages 58, 59). (The traditional Wintu word for giant man is *Supchet*.) Story of Wineepoko and Supchet.

(...) This grouse that sits on limb of fir tree is <u>a magic grouse put there by *Supchet* to fool Indian people</u>. (*Supchet* says to the human hunter Wineepoko's son):"Have you shot at my dear pet grouse? One is sitting there on limb on tree." Wineepoko's son said, "yes, but I <u>could not hit it</u>." (...) *Supchet* says to Wineepoko's son, "you look young and strong, what say <u>let's wrestle</u>?" (...) But Wineepoko's son said, "no, I don't want to wrestle." (...) ...then they wrestle and fight for a while until Wineepoko's son gave out. Then *Supchet* threw him down hard on the ground, took his heart out, took it home, going west.

Original Source: *A Bag of Bones: Legends of the Wintu Indians of Northern California,* by Marcelle Masson, 1966. Happy Camp: Naturegraph Publishers, pp. 84-91.

Comments: This is a good example of homin-human competition for game in hunting. Note *Supchet's* magic ability to fool humans. Homin-human wrestling and fighting bouts are a standard feature in Eurasian folklore.

The Shoshone (pages 80-91)

Cannibal Giant (page 84)

(The Cannibal Giant caught an old woman and her granddaughter in the wood where they were gathering pitch from pine trees. The giant killed them and took them home. He ate their bodies).

The grandfather went out to look for his wife and granddaughter when they didn't come home. He found the giant's tracks and followed them to his cave where he found the giant asleep. He had his bow and arrows with him but could not kill the giant. So he shot at his penis and that is how he killed the giant. The giant is like a rock.

Original Source: *Shoshone Tales, by Anne M. Smith*, 1993. Salt Lake City: University of Utah Press, p. 37

Comments: Here we see quite recognizable and realistic details and traits, known to investigators from fieldwork and witnesses, including cases of stumbling upon sleeping homins. Native story-tellers know, just as we do, how difficult it is to kill a giant, but the method of killing described here is probably a flight of fancy.

Tso'apittse (pages 84, 85)

Tso'apittse was a rocky giant with pitchy hands. When children are naughty their parents tell them the giant will come down from the mountain. (...) (The giant killed and ate a young woman. Her husband told his father what had happened. He and his father started making lots of arrows). His father said, "You can't hit these *Tso'apittses, their bodies are made of rock. Their only vulnerable place is in the anus*. (...) When his anus is exposed, shoot at it." (...) Then the young man took the arrow with the obsidian point and shot and it hit *Tso'apittse's* anus. (...) The young man watched *Tso'apittse* squirm and die."

Original Source: *Shoshone Tales, by Anne M. Smith*, 1993. Salt Lake City: University of Utah Press, pp. 62–65.

More on the same theme... (page 87)

When *Tso'apittse* comes, the children are held there by some power and the parents get away alone.

Comments: Sasquatches "made people sleep and took their fish away." "Henry Allen had heard they could 'make people crazy' but did not know how this was done." (SLS, pp. 57, 61).

Jarbridge (white version of Shoshone word *Tso'apittse*) (page 91)

The giant preys on Indians, tossing them into a basket slung across his back. (...) *Tso'apittse* is a hairy devil.

Original Source: Unknown

Comments: A basket slung across the giant's back is mentioned in many tales, so I wonder if it's an imaginary detail, for it is never reported by present-day eyewitnesses. The nearest thing we have from Russia is mention of "a box made of birchbark" from which two hairy human-like figures took out and ate something (probably berries), as observed by a witness (FRS, p. 181)

The Comanche (page 102)

Piamupits or *Mu pitz* is a cannibal monster who was a terrifying cave dwelling ogre, about 12 feet tall and covered in hair. Sanapia, a Comanche medicine woman, described *Mu pitz* as a very tall, hairy giant with big feet. He is huge and has a foul smell. (...) Comanche elders put out food for the *Mu pitz* because he still roams Oklahoma. Comanche grind *Mu pitz* bones into a powder and use it to treat sprains and bone problems. They tested the bones first to see if it had special power by putting the bone on their tongues.

Original Source: *Fossil Legends of the First Americans*, by Adrienne Mayor, 2005. New Jersey: Princeton University Press, pp. 196–197.

Comments: Note that the elders "put out food" for the homins (I'll touch upon this later). That powdered *Mu pitz* bones were used as medicine corresponds to similar practices in Tibetan medicine concerning other parts of the "wild man" body: "His meat may be eaten to treat mental diseases and his gall cures jaundice" (FRS, p. 67). If these beings are imaginary, then their bones, meat and galls are imaginary, too. Wish our American colleagues would attempt to learn more about those powdered bones, and, best of all, get hold of them.

Peah-Moopitz Kahnik (or *Pia Moopitsi Kahni*) (page 103, 104)

A giant lived in a cave located on the southern slope of Elk Mountain in the early days before the white man came. Exacting two buffaloes every fortnight from the Indians living south of the mountain, he was a constant and fearful menace. As the years went by and the buffalo became more scarce because of the frequent buffalo hunts of the many different tribes of Indians, the fulfilling of the giant's request was made increasingly difficult. Slowly the white men came in. They also organized buffalo hunts.

Finally, when the Indians found it almost impossible to furnish the required number of buffalo, they held a council. A young brave was designated to confer with the giant concerning their problem. Cattle were to be suggested as a substitute.

Approaching the entrance of the cave, the brave called, "Great Giant, I come before you to ask an important question."

"What is it you want?" said the giant.

"There are not enough buffalo on the prairies or in the mountains. Will you accept the beef of cattle instead? We have been eating it for years and find it very delicious."

"Cattle are very small, but I shall be satisfied if your tribe will bring me twenty," replied the giant.

Cattle were thus substituted for buffalo, but the change of the diet did not agree with the giant. (...) The coming of so many white men, bringing confusion to the quiet moun-

<u>tain country, was also disquieting to the giant</u>. The Indians, trying to appease his wrath, brought him forty beeves.

Finding his new diet more and more disagreeable, and <u>the encroachment of the white man unbearable, the giant left his cave for a more secluded spot in the larger mountain range farther west</u>.

Original Source: Daniel A. Becker. May 1940.

Comments: This is one of the most important items of information in North American hominology. First of all, note that the legend is likely very old. For certain it was created long before in-depth research in this field began, so no bias in favor of hominology could have appeared. Second, the tale sheds light on the rules and history of homin-human interactions. On the one hand, we learn of competition and fierce struggle between these two types of hominids; on the other, of deification and worship of "wildmen." How to explain this contradiction? I believe these differences refer to different historical epochs and different environments. *Homo sapiens,* while building civilization and becoming civilized, needed plains for agriculture and animal husbandry, thus pressing and driving homins from fertile plains into marshy lands, forested mountains, or deserts. As sapiens population increased, humans began to encroach on homin territories again, but in these environments those who had evolved into "wildmen" and "forestmen" had an edge over humans. Thus began worship and deification. A third stage in relationship came when heathen cults began to be replaced by monotheistic religions, and the wild, hairy giant was declared to be a devil. Here is an extract from my previous book, BRV, p. 30, partly touching on this theme:

Human-Demon Interactions

In heathen times, the demons were not devils and goblins but "gods" and "lords of nature." People worshipped them not out of superstition but for quite sensible and pragmatic reasons. Going to hunt or to fish they entered the territories of those wild hairy giants, and seeking a *modus vivendi* with them, people had to sacrifice a part of their trophies and catches to the homins. That is the origin of religious sacrifices, whose echoes are still reverberating in folklore. *(Further information is found on pages 27 and 28 in this volume.)*

The Woman Who Married a Giant (page 104, 105)

[The giant kidnapped her] <u>She thought over many plans to escape and make her way back to her camp</u> [and succeeded only thanks to the help of Brother Bullfrog and Brother Crane].

Original Source: *The North American Indian,* Vol. 19, by Edward S. Curtis, 1930. Massachusetts: The Plimpton Press, p. 196.

The Mosopelea (page 106)

The traditional Mosopelea name for bigfoot is <u>Yeahoh</u>, which means "monster" and is <u>directly related to the sound the animal made</u>.

The *Yeahoh* (page 106)

Once they was man out huntin', he got lost, and after a while he begin to get hungry. He come to a <u>big hole in the ground</u> and he thought he would venture down into it. He went down in there and he found that the old *Yeahoh* lived in there and had <u>deer meat</u> hangin' up and other foods piled around the walls. The man was afraid at first, but *Yeahoh* didn't bother him (...) and said, *"Yeahoh, Yeahoh,"* a time or two. He cut it off a peice of the meat and it started eatin' it. (...) Well, the man lived there with it a long time and they got along all right. After so long they was a young'un born to 'em, and it was <u>half-man and half-Yeahoh</u>. And the *Yeahoh* took such a liking to the man it wouldn't let him leave. He got to wanting to get away and go back home. [After an unsuccessful attempt, the man made his escape.] This time he got to the shore where there was a ship ready to set sail. He got on this ship and he looked and saw the Yeahoh comin' with the young'un. It screamed and hollored for him to come back and when it saw he wasn't goin' to come, why, it just tore the baby in two and held it out one-half to him and said, "Yeahoh, Yeahoh." He sailed on off and left it standing there.

Original Source: Interview of Lee Maggard, Putney, Harlan County, Kentucky, 1950. *Western Folklore-Vol. XVI,* January 1957, No.1.

Comments: Two things are striking and amazing here. One is that *"Yahoo"* which seems a variation of *"Yeahoh,"* is applied to wild, hairy bipeds by Jonathan Swift in *Gulliver's Travels* and by the inhabitants of Australia. Second, the tale, with some local variations of the plot but with the obligatory tearing of the baby "in two" at the end, is truly world-wide. I've read several such accounts from different parts of the globe, including this country, and wonder how such international "consensus" in folklore could have come about.

The Sioux (page 108)

The traditional Sioux words for a bigfoot-like creature are *Chiye-tanka* (big man) and *Iktomi* ("The Trickster" or "Double Face").

Comments: These creatures are "fond of playing tricks on humans, such as sneaking up and kicking them, tying them to trees with thongs lashed to the genitals, etc.," (SLS, p. 62). Demons in Russia also love dancing and merrymaking, especially all kinds of pranks, so that Russian peasants called them "jokesters" and "pranksters." A favorite prank of rusalkas was to catch wild geese on the river and entangle the feathers of their wings so that the birds could not fly. Or they would let the fish out of the fishermen's net and fill the latter with slime and water-plants, or divert themselves by putting out a fishermen's or hunter's campfire with the water dripping from their hair covering. (BRRV, p 30). Janice Carter Coy knows well from experience bigfoot tricks and pranks performed on her farm.

Chiye-tanka (page 108)

Chiye-tanka or Big Man is a kind of husband of *Unk-ksa*, the Earth, who is wise in the way of anything with its own natural wisdom. Sometimes we say that this one is a kind of animal from the ancient times who can take a big hairy form; (...). The Big Man comes from God. He's our big brother, kind of looks out for us.

Original Source: *In the Spirit of Crazy Horse,* by Peter Matthiessen, 1980. Minnesota: Viking Press.

The Ojibwa (page 128)

The traditional Ojibwa name for a wild man is *Puck Wudj Ininees.*

Puck Wudj Ininees (pages 128, 129)

And I shall ever be called *Puck Wudj Ininees,* or the little wild man of the mountains.

Original Source: *North American Indian Legends,* by Allan A. Macfarlan, 1968. New York: Dover Publications, pp. 71–75.

Commentary: The first story in Kathy Moskowitz Strain's book, beginning on page 15, is titled "The Cannibal Dwarfs." Here are some quotes from other books:

> I wonder if you have ever heard of the Little Red Men of the Delta? (...) They are said to be about the size of a ten year old kid and able to climb like monkeys and to live back from the bayous. They talk a lot but keep out of gunshot range and mostly go into the water. They are people and the muskrat trappers say they often wear scraps of discarded lines (linens?) old jeans and such"[compare with Russian homins – D.B.] *(Abominable Snowmen: Legend Come To Life,* Ivan Sanderson, 2006, p. 96).

...the earth <u>dwarfs</u> ("little earths") who lived in nooks, crannies, and forest recesses and could control the game and steal human souls (SLS, p. 61).

Little people in the Indian culture live in the hills surrounding the reservation. They are extremely rare, hardly ever seen and are considered sacred. They are normally not over four feet tall, but sometimes can grow to the height of a normal person. (...) I have personally met and interviewed several people who claim to have seen and interacted with the little people (THP, p. 222).

It appears that *El Duende,* according to these people, is a <u>dwarf</u> who lives in deep canyons and desolate valleys [in the Colombian Andes, South America – D.B.], where he can often be heard crying like a baby or, when he is in a boisterous mood, making noises rivaling thunder. Natives firmly believe that he is very fond of horseback riding, but being so small, is unable to sit on the horse's back, so he sits on the animal's neck, making stirrups by plaiting the mane in such a way as to be able to put his feet in it (*Tschiffely's Ride,* Aimé F. Tschiffely, 1933, New York, p. 182).

...soon afterwards we began to hear bits and pieces of Aboriginal lore that seemed to refer to a widespread belief in the existence of similar <u>tiny hairy men</u> in various parts of Australia. (*The Yowie: In Search of Australia's Bigfoot,* Tony Healy and Paul Cropper, 2006, Sydney, Australia, Strange Nation, p. 121).

Wild, hairy bipeds, both giants and dwarfs, are reported in the Caucasus. The latter have also been observed in Africa. Aside from the Orang Pendek investigation, pygmy homins remain a mostly unexplored subject of hominology.

The Shawnee (page 131)

The Shawnee were original residents of Ohio, Kentucky, and Pennsylvania. The current population resides in Oklahoma, Alabama, and Ohio. The Shawnee language is in the Algonquian stock.

The Hairy Woman (page 131)

One time I's prowling in the wilderness, wandering about, kindly got lost and so weak and hungry I couldn't go. When it began to get cool, I found a big cave and crawled back in there to get warm. Crawled back in and come upon a leaf bed and I dozed off to sleep. I heard an awful racket coming into that cave, and something come in and crawled right over me and laid down like a big old bear. It was a hairy thing and when it laid down it went chomp, chomp, chawing on something. I thought to myself, "I'll see what it is and find out what it is eating."

I reached over and a hairy like woman was there eating chestnuts, had about a half a bushel there [How did she carry them?– D.B.]. I got me a big handful of them and went to chewing on them too. Well, in a few minutes she handed me over another big handful, and I eat chestnuts until I was kindly full and wasn't hungry any more. D'rectly she got up and took off and out of sight.

Well, I stayed on there till next morning and she come in with a young deer. Brought it in and with her big long fingernails she ripped its hide and skinned it, and then she sliced the good lean meat and handed me a bite to eat. I kindly slipped it behind me, afraid to eat it raw and afraid not to eat it being she give it to me. She'd cut off big pieces of deer meat and eat it raw. Well, I laid back and the other pieces she give over as she eat her'n. She was goin' to see I didn't starve.

When she got gone again I built me up a little fire and br'iled my meat. After being hungry for two or three days, it was good cooked-yes, buddy. She come in while I had my fire built br'illing my meat, and she run right into that fire. She couldn't understand because it kindly burnt her a little. She jumped back and looked at me like she was going to run through me. I said, "Uh-oh, I'm going to get in trouble now."

Well, it was cold and bad out, so I just stayed another night with her. She was a woman but was right hairy all over. After several days I learnt her how to br'ile meat and that fire would burn her. She got shy of the fire and got so she liked br'iled meat and wouldn't eat it raw any more. We went on through the winter that way. She <u>would go out and carry in deer and bear</u>. So <u>I lived there about two year</u>, and <u>when we had a little kid, one side of it was hairy and the other side was slick</u>.

I took a notion I would leave there and go back home. I begin to build me a boat to go away across the lake in. One time after I had left, I took a notion I would slip back and see what she was doing. I went out to the edge of the cliff and looked down into the mountain, and it looked like two or three dozen of hairy people coming up the hill. They were all pressing her and she would push them back. They wanted to come on up and come in. I was scared to death, afraid they's going to kill me. <u>She made them go back and wouldn't let them come up and interfere</u>.

Well, I took a notion to leave one day when my boat was ready. I <u>told her</u> one day I was going to leave. She follered me down to my boat and watched me get ready to go away. She <u>was crying, wanting me to stay</u>. I said, "No, I'm tired of the jungles. I'm going back to civilization again, going back."

When she knowed she wasn't going to keep me there, she just grabbed the little young'un and tore it right open with her nails. <u>Throwed me the hairy part and she kept the slick side</u>. That's the end of that story.

Original Source: Interview of Joe Couch, Appalachia, Virginia, 1954. *Western Folklore-Volume XVI*, January 1957, No. 1.

Comments: I see this as a traditional story pretending to be a factual testimony, which impressed its listeners with realistic and valuable details. Of interest is the difference in their attitude toward fire. Marie-Jeanne Koffmann recorded in the Caucasus similar stories of almasty entering and sharing a cave with a human stranger.

Bella Coola *Boqs* (bush man) (page 136)

[The *boqs*] somewhat resembles a man, its hands especially, and the region around the eyes being distinctly human. It walks on its hind legs, <u>in a stooping posture</u>, its long arms swinging below the knees; <u>in height it is rather less than the</u>

average man. The entire body, except the face, is covered with long hair, the growth being most profuse on the chest which is large, corresponding to the great strength of the animal. (...) It is said that a woman was once drawing water at the edge of a stream when a boqs, concealed on the other shore, extended its penis under the water to the further bank and had intercourse with her. The contact rendered her powerless, as if turned to stone; she could neither flee nor remove the organ. Her companions tried unsuccessfully to cut the organ until one of them brought a salalberry leaf, whereupon the monster, dreading its razor-like edge, withdrew.

Original Source: *Legends Beyond Psychology,* by Henry James Franzoni III and Kyle Mizokami.

Comments: A marvelous super-hyperbole on a fully realistic theme of homin-human relationship. I have covered this topic on pages 32 to 34 of this volume, and provide here additional information from my book, BRV, pp.35–36:

The Sexual Connection

(...) In Asia, the 12th century Persian scholar Nizami al-'Arudi wrote that "the Nasnas, a creature inhabiting the plains of Turkestan, of erect carriage and vertical stature, [...] is very curious about man. [...] And if it sees a lonely man it abducts him and is said to be able to conceive by him. This, after mankind, is the highest of animals..." Modern scholars say the Nasnas is an imaginary creature, a kind of faun.

Sexual relations with demons is a topic present in all works on folklore that I read and referred to in my book. In Tajik folklore, the female demon "pari" seeks the love of a hunter and pays him with wild goats that she sends him in gratitude.

In Chuvash folklore, the female arsuri (goblin dubbed monkey) would run in the woods in front of a man, laughing impudently, showing him her genitalia and beckoning to him. The name "arsuri" is applied by the Chuvash to a shameless woman.

In Circassian folklore it is said that the shaitan and his female partner jinne can be caught. However it is not advisable to catch a shaitan because he will offer strong resistance. Jinne is a different matter. If caught, she can be used as a woman. Sometimes she herself is seeking sex with humans, coming to herdsmen for the purpose.

The Chehalis (page 140)

The Chehalis (also known as the Sts'ailes) occupy an area near the Fraser River in their homeland of <u>British Columbia</u>. Their traditional word for a bigfoot-like creature is *Saskehavis*, meaning "wild man." <u>John W. Burns, a teacher for the Chehalis reserve in Harrison Mills from 1925-1945, coined the word "Sasquatch" based on the various names used by tribes within the Salishan language group</u>. In 1980, the Chehalis band in British Columbia adopted a Sasquatch image as their symbol.

What happened to Chehalis Native Serephine Long? In her own words, here is her intriguing story.

I was walking toward home one day, many years ago, carrying a big bundle of cedar-roots and thinking of the young brave Qualac (Thunderbolt), I was soon to marry. Suddenly, at a place where the bush grew close and thick beside the trail, a long arm shot out and a big, hairy hand was pressed over my mouth. Then I was suddenly lifted up into the arms of a young "Sasquatch." I was terrified and fought, and struggled with all my might; in those days I was strong. But it was no good; the "wild man" was as powerful as a young bear. Holding me easily under one arm, with his other hand he smeared tree-gum over my eyes, sticking them shut so

Chehalis logo. It was created by Ron Austin, a Chehalis Native, in 1980.

that I could not see where he was taking me. He then lifted me to his shoulder and started to run.

He ran on and on for a long long time—up and down hills, through thick brush, across many streams, never stopping to rest. Once he had to swim a river, and then perhaps I could have gotten away, but I was so afraid of being drowned that I held on tightly with my arms about his neck. Although I was so frightened I could not but admire his easy breathing, his great strength and his speed of foot.

The only known photograph of Serephine Long. It is believed she was about 87 when the photo was taken. (Photo: Public domain)

After reaching the other side of the river, he began to climb and climb. Presently the air became very cold. I could not see, but I guessed that we were close to the top of a mountain.

At last the "Sasquatch" stopped hurrying; then he stooped over and moved slowly, as if feeling his way along a tunnel. Presently he laid me down very gently, and I heard people talking in a strange tongue I could not understand. The young giant next wiped the sticky tree-gum from my eyelids and I was able to look round me.

I sat up and saw that I was in a great big cave. The floor was covered with animal skins, soft to the touch and better preserved that we preserve them. A small fire in the middle of the floor gave all the light there was.

As my eyes became accustomed to the gloom I saw that beside the young giant who had brought me to the cave there were two other "wild people"—a man and a woman. To me, a young girl, they seemed very, very old, but they were active and friendly, and later I learned that they were the parents of the young "Sasquatch" who had stolen me. When they all came over to look at me I cried and asked them to let me go. They just smiled and shook their heads.

From then on I was kept a close prisoner; not once would they let me go out of the cave. Always one of them stayed with me when the other two were away.

They fed me well on roots, fish and meat. After I had learned a few words of their tongue, which is not unlike the Douglas dialect, I asked the young giant how he caught and killed the deer, mountain goats, and sheep that he often brought into the cave. He smiled, opening and closing his big, hairy hands. I guessed that he just laid in wait, and when an animal got close enough, he leaped, caught it, and choked it to death. He was certainly big enough, quick enough and strong enough to do so.

When I had been in the cave for about a year I began to feel very sick and weak and could not eat much. I told this to the young "Sasquatch" and pleaded with him to take me back to my own people. At first he got very angry, as did his father and mother, but I kept on pleading with them, telling them that I wished to see my own people again before I died.

I really was very ill, and I suppose they could see that for themselves, because one day, after I cried for a long time, the young "Sasquatch" went outside and returned with a leaf full of tree-gum. With this he stuck down my eyelids as he had done before. Then he again lifted me to his big shoulder.

The return journey was like a very bad dream, for I was light-headed and in much pain. When we recrossed the wide river I was almost swept away; I was too weak to cling to the young "Sasquatch." But he held me with one big hand and swam with the other.

Close to my home he put me down and gently removed the tree gum from my eyelids. When he saw that I could see again he shook his head sadly, pointed to my house and then turned back into the forest.

My people were all wildly excited when I stumbled into the house, for they had long ago given me up as dead. But I was too sick and weak to talk. I just managed to crawl into a bed, and that night I gave birth to a child. The little one lived only a few hours, for which I have always been thankful. I hope that never again shall I see a "Sasquatch."

Source: "The 'Hairy Giants' of British Columbia," by J.W. Burns, set down by C. V. Tench, 1940. *The Wide World,* January, Vol. 84, No. 502.

Comments: This story is absolutely sensational! Who is Serephine Long? Not a word of explanation in the magazine article. If she is the heroine of the story, she must not be less (probably more) famous than Albert Ostman—unless, of course, Serephine's story is not true. Albert Ostman was interviewed by John Green and René Dahinden, grilled by newsmen, cross-examined by a magistrate, a zoologist, a physical anthropologist and a veterinarian. Who interviewed Serephine Long other than J. W. Burns? The name Albert Ostman is known to every hominologist worth his salt. It does not appear that many people have heard of Serephine Long. Chris Murphy appears to be the first, after Tench, to mention her name in print—his book, *Meet the Sasquatch,* 2004. He cites John Burns:

> But perhaps the strangest experience happened to a Chehalis woman, Serephine Long. She told me she was abducted by a Sasquatch and lived in the haunts of the wild people for about a year.

Another quote:

> I've never personally encountered a Sasquatch myself. Yet I've compiled an imposing dossier of first-hand accounts from Indians who have met the wild giants face to face and know survivors of the tribe still live today. I was always aware when the Sasquatch were in the vicinity of our Indian village, for then the children were kept indoors and not allowed to venture to my school. The Chehalis Indians are intelligent, but unimginative, folk. Inventing so many factually detailed stories concerning their adventures with the giants would be quite beyond their powers.

This is also one of my own criteria in assessing eyewitness accounts. One more quote: "Many of my other Indians (besides Serephine Long – D.B.) are sincerely convinced the Sasquatch live in the unexplored interior of B.C. And with the Indians, whom I know and trust, I also believe." (*Meet the Sasquatch*, pp. 31, 33).

Why is it that his crucial information is coming to light only now, and only through interference from abroad? My explanation concerns John Burns' reputation with some veterans of the bigfoot investigation in North America. Burns not only coined the word "Sasquatch"—he equated the latter with humans, not apes, and for this reason has been ignored by most bigfooters.* That sasquatch, at least some of them, are people of subtle and well-developed mind is apparent from Serephine Long's account more than from anything else. Assuming the account is true, every word of it is as precious as gold. For the moment, I opt to single out only her telling that sasquatch have a language. This supports the groundbreaking claims by Albert Ostman, Janice Carter Coy and Scott Nelson. I can also add from Wayne Suttles: "They called these people 'wild tribes' who traveled by night and attacked lone wayfarers. (...) They spoke a language unintelligible to the Snohomish." Also that the giants can "talk like an eagle, owl, screech owl, and bluejay." (SLS pp. 58, 57).

John W. Burns in 1946; an ea[rly] hominologist.

And one more extract from my book may be appropriate:

Editors' Note: It is not so much that Burns was "ignored" as it is the early timing of the release of his known published findings (1929, 1940, and 1954). Although John Green re-published Burns' first known article (1929) in his *The Sasquatch File* (1973), the story of Serephine Long is not in this article.

Crossbreeding

The basic difference of demons from all real creatures, including apes and monkeys, is their desire for sexual relations with man. Clearly, this circumstance is responsible for their unprecedented and unique role in the history of mankind. A Russian specialist on oriental folklore and the Koran wrote in 1893 about the demons called "jinn": "The peculiarity of their nature is that they can have sexual intercourse with people."

A natural question then is: What comes as a result of such intercourse? Folklore is quite talkative on this score. An item from Siberia: "Sometimes a she-devil cohabits with hunters in the forest and becomes pregnant from them, but she tears the infant apart at its very birth." The Circassian jinne can also kill her crossbreed baby, in case her human husband reveals her presence to his neighbors.

A success story in crossbreeding is reported by Kazakh folklore, telling of a horse herdsman who encountered a female almasty in the steppe and thought, "Be it a shaitan or a human, it doesn't matter." He lived with her and "they had three children born to them."

Bashkir folklore explains the origin of the name of the Shaitan-Kudey clan by the fact that once a brave Bashkir caught and married a female shaitan and their posterity formed the said clan.

Nogai folklore notes the rapid growth and unusual strength of the offspring of their legendary hunter Kutlukai and his almasty wife. Their son became a national hero and all Nogai nobility descend from him.

If we give credence in this respect to folklore, then hominology is faced with the question: What is the genetic status of "demons," i.e., homins, in relation to *Homo sapiens?*

"Good" species are not supposed to produce fertile crossbreeds. Still, division into species and subspecies of closely related organisms is often a matter of speculation and agreement.

Primatologists are aware of fertile hybrids of different

monkey species. Another case in point is the example of wolves and coyotes, considered to be different species. Yet they carry the same number of chromosomes and there exist no genetic barriers to their interbreeding. If not for behavioral differences, which keep them separate, one species would have long ago absorbed the other.

The homin-human situation appears to be similar; the barrier to crossbreeding is neurological and behavioral, not genetic. For these reasons it can be overcome in principle and in practice, but the process has been "invisible" and very protracted.

One more example in favor of this view is a quote from *Essays on Russian Mythology* (1916) by D. K. Zelenin: "People believe that if a rusalka is made to wear the cross, she will become a human being. Such cases are reported from the Vladimir Province, where two boys married baptized rusalkas."

As regards North America, Dr. Ed Fusch reports crossbreeds between Indians and the "Stick Indians" (Sasquatch, "Night People") in *S'cwen'yti and the Stick Indians of the Colvilles* (1992). (Posted by the late Bobbie Short on her Bigfoot Encounters Internet site and supplied to me by the late Don Davis.) (BRV, pp.37, 38).

I have covered only half of Kathy Strain's book, so it's too early to come to final conclusions, but one or two preliminary ones should be stated. I hope it's clear to the reader by now how important the knowledge of relevant folklore is for our field of study. North American hominologists have set some world records in practical terms; the Patterson/Gimlin documentary film. the number of footprint photos and casts, and the priceless Carter Farm evidence. But in terms of theory, in a theoretical vision and understanding of the phenomenon, Russian hominologists are ahead. And this is because folklore and demonology have been regarded by Boris Porshnev and myself, in his footsteps, as a highly valuable source of evidence from the very beginning. Had Kathy Strain's book been in existence 30 years ago, the situation, in terms of theory, could be different today in North America as well.

The Chinook (page 142)

The Chinook have <u>several names</u> for a bigfoot-type monster, <u>depending on gender and location</u>. The most famous, *Skookum*, is translated as <u>"Evil God of the Woods"</u> or <u>mountain devil</u>. Other terms include (...) *Itohiul* (<u>big feet</u>).

Comments: So the name BIGFOOT was first coined by Native Americans and much earlier than the 1950s!

Kihlktagwah, the *Itohiul* (page 148)
There was an *itohiul* who could <u>walk</u> across the river <u>on the water</u>.

Original Source: *The North American Indian,* Vol. 8, by Edward S. Curtis, 1930. Massachusetts: The Plimpton Press, pp. 15–153.

Unseen Bigfoots (page 152)

(...) The Clackamas Indians maintain that in the lands of the headwaters of the Clackamas River, adolescent Bigfeet beings have to pass <u>a test</u> to become an adult member of the Bigfeet tribe. They must jump in front of a human on a trail, and wave their hands in front of the human's face, <u>without being seen</u>.

Original Source: *Attitudes Toward Bigfoot in Many North American Cultures,* by Gayle Highpine, 1992. In: The Track Record #18.

Comments: The above is an explanation of the famous sasquatch "invisibility" without resort to the theory of "other dimensions." If the alleged test is true, it seems to be only possible thanks to homin paranormal abilities.

The Kwakiutl (page 153)

The Kwakiutl were made up of several tribes that occupied the Northwest Coast and whose traditional languages were in the Wakashan language group. The tribes are better known today as the Kwakwaka'wakw and primarily occupy north Vancouver Island and British Columbia. The Kwakwaka'wakw have five different names for bigfoot, which depends on gender. These include *Be'a'-nu'mbe* ("Brother of the Woods"), *Bukwas* ("Wild man of the Woods"), *Dzunukwa* (*Dsonoqua*) ("Wild Woman of the Woods"), *Tsonaqua* ("Wild Woman of the Woods"), and *Tsunukwa* ("Female giant covered with hair with big feet"). It should be noted that *Bukwas* is the son of *Dzunukwa* and a human male, as told in the first story.

Dzunukwa (page 153)

Dzunukwa (or *Dzoonokwa, Dzonoqua*, or *Tsonoqua*) is a fearsome giantess of the dark forest that is not-quite-human female. She is also known as Wild Woman of the Woods or Property Woman. She has black hair, pendulous breasts, heavy eyebrows, deep-set eye sockets with half-closed eyes and has pursed lips...she cries "Ooh-ooh, ooh-ooh." She is stupid, clumsy and sleepy. She captures children who are crying and who venture into the forest, carrying them away in a basket on her back to devour them. Her house is filled with wonderful treasures such as boxes of food, coppers, canoes, and more. Through special encounters with her, a person can acquire some of the wealth and supernatural powers.

One day she stole some dried fish from a Kwakiutl man. The man pursued her and caught her. They became lovers and produced a son, *Bukwas*. One day a young man found her baby in its cradle in the forest. He teased the baby by pinching it, causing it to cry loudly. *Dzunukwa* heard the cry and called out, "Whoever you are that may be teasing my baby, let him alone and I will give you a spear."

Pleased at such good fortune, the young man pinched the baby three more times and was offered the Water of Life, a magic wand, and a supernatural canoe if he would leave the baby alone. Satisfied, the man stopped teasing the child, returned home with his gifts and, because of his encounter with *Dzunukwa*, became rich and powerful.

Original Source: Unknown

Comments: Folklore doesn't fail to mention distinctive anatomical features of the "mountain devils," including even such paleoanthropologically significant traits as heavy eyebrows and deep-set eye sockets—characteristic of pre-sapiens hominids. As we understand now, mention of "supernatural powers" is not due to fancy but to the homins' real paranormal powers. "The Water of Life" is also present in many Russian folk tales.

Dzunukwa and *Bukwas* (page 154)

When *Dzunukwa* steals a female child, she keeps it as her daughter and picks salmon berries for her. She also likes to steal salmon from the village. She throws aside the roof boards and reaches down to take the fish from the drying frames. Her son *Bukwas* is in the habit of striking trees with a piece of wood. His body is hairy and he is shy.

Original Source: *Bella Bella Tales* by Franz Boaz, 1932. New York: The American Folklore Society.

Comments: Very realistic.

Thunderbird and *Tsonoqua (page 154)*

Chief Splashing-Waters was having difficulties with the Wild Woman of the Woods. Thunderbird, *Kwun-kwane-kulegui*, came to his rescue and turned the savage *Tsonoqua* into stone. In remembrance of this help, the Chief decreed that Thunderbird would be respected as the Protector of Man and as the Spirit that made wishes come true. *Tsonoqua* was placed under him, to be ruled by him, and is why he is often shown in totem poles with him sitting on the savage's head. Songs, dances, and masks were made to honor *Kwun-kwane-kulegui*. *Tsonoqua* would now forever be represented with spouting lips, symbolizing that she blows the wind in the forest.

Original Source: Unknown.

Comments: I think the above can be regarded as evidence of the overcoming by the Indians of homin cults as supreme cults and regarding them as second to the spiritual heaven-located (<u>Thunder</u>bird) deity, an epoch-making change that is typical of human-homin historical relations elsewhere in the world.

The First *Tsunukwa* Dance (page 160)

(...) ...a strange creature with <u>great, hanging breasts</u>, and a round, <u>protruding mouth</u>. (...) Its <u>eyes are enormous</u>, and there seems to be <u>fire inside them</u>.

Original Source: *The North American Indian,* Vol. 10, by Edward S. Curtis, 1930. Massachusetts: The Plimpton Press, p. 296–297.

Comments: The reader will recall the Sioux saying "The Trickster" is "a kind of animal from the ancient times." Hominids from the ancient times, i.e., preceding sapiens, had "protruding mouths." Eye-sockets of Neanderthal skulls are larger than those of *Homo sapiens.* That homin eye-shine can be seen at night, like those of animals with night vision, is well-known, but that their eyes can be internally illuminated (have "fire inside them") I could not accept for a long time. Today, my friend and colleague, Dr. Michael Trachtengerts, a physicist, is supporting and developing this theory.

Big Figure (page 162)

It was a <u>big, big man, bigger than any other</u>. <u>He had hair all over his body</u> and <u>his eyes were set deep in his face</u>. He carried a large basket on his back.(...) Then he said to the men, "Why are your faces so nice and smooth and <u>not rough like mine</u>? You have nice eyes. <u>They don't sink way in your head like mine do</u>."

Original Source: *Kwakiutl Legends,* by James Wallas and Pamela Whitaker, 1989. Surrey, BC: Hancock House, pp. 154–157.

Comments: Another mention of distinctive anatomical features.

Big Figure's Wife (page 165)

(...) Finally, the hunter came around to the front of the woods giant's woman.
"Yes, it is I," he said. "What are you making the canoe for?"
"We live beside a long lake," said the giantess. "We will use it there.
Why have you come to me?" she asked him.
"I followed the sound of your axe," he replied, "and now I have found you I want something from you."
"What is it that you want?"
"I am a provider of food for my people and I have not had much luck lately in hunting. Can you help me?"
"I will help you," responded the lady. "I will use my power to bring elk, deer and bear to you. When you are hunting in the water, seal will come to you."
The hunter was pleased at the big woman's generosity, yet he asked her for one more favor. "I want to use your features in a dance mask," he said.
"If you use me, you must use all of me and my four children too," she replied. "This baby is the youngest of the four. You may use us all in a dance."
After that the hunter became a very successful provider of food, and a dance was created showing the huge woman with her four babies being born one by one.

Original Source: *Kwakiutl Legends,* by James Wallas and Pamela Whitaker, 1989. Surrey, BC: Hancock House, pp. 162–164.

Comments: I wonder if the giantess really used an axe. Janice Carter Coy's bigfoots used hands and sticks to dig a grave while they had easy access to spades.
Note that the images of satyrs, nymphs, and sileni, with appropriate masks, played permanent roles in ancient Greece during the birth of the European theater.
That homins may help friendly humans in hunting is well-known, and not only from folklore but past reports as well. Below is the final part of a seemingly truthful story which was sent to us from America and is now posted, with the informant's permission,

on Michael Trachtengerts' website (www.alamas.ru). The events described therein happened in 1994, in the Blue Mountains of the state of Washington.

Hunting Trip, Day 5—Departure: Early the fifth morning (actually the sixth day we were out, since we had driven the first day), all the fruit and sandwiches [put out as offerings to bigfoot] were gone. Obviously, the bigfoots had come in the night and taken our offerings. Pop [Indian hunting guide] insisted that we scratch out the bigfoot footprints nearby, and by the horses, so no one else would know that there were bigfoot living in the area, in case hikers might come across Pop's hunting camp. So we dragged some branches over the bigfoot footprints. Pop felt strongly he should <u>protect their territorial rights</u>. He had a lot of respect for the bigfoot. He thought of them as <u>a special kind of people</u>.

We cleaned the camp, put the horses in the horse trailers, and left the camp in good order. This was <u>Pop's camp. He depended on it for part of his living</u>. He was really good at what he did. He is deceased now; and I remember him as a good man!

Aside: *The narrative of the hunting trip continues, and we come to the point where three hunters (including Pop) are hunting on foot about 4 miles from their camp. One of the hunters brings down a very large elk. However, it is apparently too late in the day to go back to camp to get horses to drag the carcass out. All they can do is gut it and come back the next day. Pop, who once had a similar situation (explained later), leaves an offering of sandwiches by the carcass. The men return to their camp and in the morning find the elk carcass in the vicinity of their camp. The hunters (excluding Pop) cannot believe their eyes.*

A few days later, we heard that the hunter who had shot the giant bull elk got 1000 pounds of meat from the animal, not counting the head, or the weight of the bones! This means that <u>the bigfoots had probably carried at least 1300 pounds of elk for us, for over 4 miles</u>, as a return favor for us giving them the peanut butter and jelly sandwiches!

Back in Walla Walla, Washington, I asked Pop if he had actually left the sandwiches for the bigfoot and not for the birds and animals as he had explained to the hunters. I hadn't directly asked him this before. Pop then told me about an experience that he once had after leaving sandwiches for the bigfoot. Apparently when hunting alone, he killed and then field-dressed a large elk. Because of weight considerations, he was only able to take one-half of the butchered elk with him, on his own horse. The other half of the dressed elk, he left right there in the forest. It was also a rough area of forest and rocks, to ride in by horseback. Surprisingly, the next morning, he discovered that the other half of the dressed elk which he had left in the remote forest, had been delivered to him during the night, by the bigfoot, near to where he was sleeping!

So with this prior experience in mind, he figured that on this trip, by again making an offering-of-sandwiches, he might get some help from these forest beings. Pop told me: "I knew that the bigfoots were following us in the forest, and watching us all the time. They stay real well hidden; but I saw them. For several days this was happening. I knew it, <u>but I didn't say anything; I definitely did not want the hunters to know</u>. Not at all! So, I was just hoping for any help the bigfoot could give us with getting the elk back to our camp, if in fact we needed their help. And as it happened, we did need help! And they came through for me!"

This is what happened, just like I have said. (Underlining type is mine – D.B.)

The Nehalem (page 168)

Strongly associated with the Clatsop Tribe, the Nehalem occupied the Oregon Coast from Tillamook Head to well south of Tillamook Bay. The tribe still lives in Oregon today. Their language is part of the Penutian group. The Nehalem had two names for a bigfoot-like creature which depended on the gender of the animal. A *Qe'ku* was a wild woman while a *Yi' dyi'-tai* was a wild man.

Wild Man (page 168)

People were drying fish up the Nehalem River. They heard a noise, the brush was crackling loudly, they knew that no wind nor common animal could be making that kind of noise. They hurried into their canoes and crossed over to the other side of the river. They forgot their little dog. They crawled into a place and lay down to listen. Their little dog barked and barked, then suddenly quit. Then they heard a terrific noise as Wild Man knocked down one side of the house. Then he must have gone back into the woods. They could not sleep they were so frightened, although they knew it was such a deep river he would be unable to wade it.

The next day one fellow went over in a canoe to have a look. One side of that large house where they had dried fish was smashed to pieces. The dog was lying there dead, and Wild Man's huge tracks were all around. That fellow came back and told the people, "Yes, I saw his tracks." They put all of their belongings and their fish in canoes and left that place for good. They would not live there any more for fear he might come again. After that no one would camp on that side of the river.

That really happened.

Original Source: *Nehalem Tillamook Tales,* by Elizabeth Derr Jacobs, 2003. Corvallis: University of Oregon Books.

Comments: Very realistic. Wild Man must have protected his territory.

Wild Men (Second Entry) (page 168)

There must have been a whole tribe of Wild Men because there were always some around.

A Nehalem man was not married. He would go hunting and permit the married people to have the meat he got. One summer he killed an elk, and he saved the blood. He took the elk's bladder and filled it with the blood. He made a camp near there. He placed that bladder of blood near his feet, lay down, and went to sleep. Wild Man came and helped himself to the elk meat.

The man awoke. He was too warm, he was sweating. "Goodness! What is the matter?" he asked himself, looking about. It was like daylight, there was such <u>a great fire burning there. Wild Man had placed large pieces of bark between the man and the fire so the man would not get too hot while he slept.</u> You see, he <u>treated that fellow well. When he spoke to him, Wild Man called the man "My nephew."</u>

The man awoke to see Wild Man, that <u>extremely large man</u>, sitting by the fire. He had the fat ribs and front of that elk <u>on a stick, roasting them by the fire</u>. He said, "This is how I am getting to be. I am getting to be always on the bum, these days. I travel all over, I cannot find any elk. <u>I took your elk, dear nephew</u>, I took your elk meat."

That man stretched himself, he had forgotten about that bladder of blood. He kicked it with his feet, causing it to make a noise. Wild Man looked around; he said, "It sounds as if a storm were coming." (<u>A Wild Man does not like to travel when it is storming</u>.) Wild Man was afraid of that noise, he kept kicking that bladder of blood. He said, "Yes, a storm is coming." Wild Man asked, "My dear nephew, would you tell me the best place to run to?" That man showed Wild Man a <u>high bluff</u>. "Over in that direction is a good place to run," he told him. Wild Man started out running. <u>Soon the man heard him fall over that bluff.</u>

The man did not go back to sleep any more that night. In the morning he went to look. There Wild Man lay, far down at the foot of the bluff. He went around by a better route and climbed down to see the body. He took Wild Man's quiver, he left Wild Man lying there. Then he became afraid, so he made ready and returned from the woods taking as much meat as he could carry. He said, "Wild Man found me. He jumped over the bluff." He too found <u>all kinds of bones in that quiver</u>.

They must have been lucky pieces because elk would come down from the mountain for him, and only he could get sea lions on the rocks.

Original Source: *Nehalem Tillamook Tales,* by Elizabeth Derr Jacobs, 2003. Corvallis: University of Oregon Books.

Comments: In this fancy tale Wild Man appears humane and the human inhumane. The question concerning whether or not homins use fire (how, where and when) remains.

The Coeur d'Alene (page 181)

(...)Their traditional Salishan word for bigfoot is not known.

Giants

Giants were formerly common in Coeur d'Alene country. They had a <u>very strong odor</u>, like the odor of burning horn. Their <u>faces were black</u>—some say they were painted black, and the giants were <u>taller than the highest tipis</u>. When they saw a single tipi or lodge in a place, they would crawl up to it, rise, and look down the smoke hole. <u>If several lodges were together, the giants were not so bold.</u>

<u>Most of them dressed in bearskins,</u> but <u>some wore other kinds of skins with the hair left on</u>. They <u>lived in caves in the rocks</u>. They had a <u>great liking for fish, and often stole fish out of people's traps</u>. Otherwise, they did not bother people much. They <u>are said to have stolen women occasionally</u> in other tribes, but there is no tradition of their having stolen women in the Coeur d'Alene country.

Original Source: *Indian Legends from the Northern Rockies,* by Ella E. Clark, 1966. Norman: University of Oklahoma Press, pp. 113–114.

Comments: That some wildmen wear clothes is reported occasionally in different areas. The subject needs more evidence, concerning animal skins in particular.

The Kootenai (page 192)

The Giant

(...) Then he [a hunter – D.B.] threw a piece of the bighorn sheep meat into the fire. When it was cooked he ate it, but it was without taste. He thought, "I'll cut a piece of my own body and I'll roast it in the fire." Then he cut a piece off of himself and threw it into the fire. When it was done he ate it. It tasted good. He cut off another piece and threw it into the fire and ate it. After two days he had devoured himself entirely. Only his bones were left. (...) He had been the first of the cannibal giants.

Original Source: Unknown.

Comments: What an interesting theory concerning possible cannibal origins!

The Modoc (page 194)

The Modoc were historic residents of northeastern California and central southern Oregon. Today, as a result of the Modoc War of 1872–1873, they are split into two major tribes, one living on a reservation in Oregon and the other in Oklahoma. Their language is part of the Penutian family. The traditional Modoc word for bigfoot is Matah Kagmi. The Modoc/Klamath traditional word is Yah'yahaas.

Matah Kagmi

[Grandfather] was walking along a deer trail near a lake just about dusk, when he saw up ahead something that looked like a tall bush. Upon coming a little closer he became aware of a strong odor, sort of musky. He then gave a close look at the bush, and suddenly realized that it was not a bush at all, for it was covered from head to foot with thick coarse hair, much like horsehair. He took a step closer, but the creature made a sound that sounded like *"Nyyaaaah!"* Grandfather now knew this was the one the old ones spoke about, a Sasquatch!
Although it was growing darker, Grandfather was able to see quite clearly two soft brown eyes through the hairy head part, then the creature moved slightly, and Grandfather made a motion of friendship and laid down the string of fish that he

had been carrying. The creature evidently understood this, as it quickly snatched up the fish and struck out through the timber nearby. It stopped only for a moment and made a sound that my grandfather never forgot—a long, low "Aaagoooooo-ouummmmmt." Grandfather called them people. He referred to them as people called "matah kagmi."

It was only a few weeks after his encounter with the *matah kagmi* that he was awakened one morning by some strange noises outside his cabin. Upon investigating, he found a stack of deerskins fresh and ready for tanning. Off in the distance he heard that strange sound once again, "Aaagooooooou-ummm!" After this there were other items left from time to time, such as wood for fuel, and wild berries and fruits.

Drawing made by the grandfather. depicts the creature when it was first seen. The inscription states, "Fourteen hand above," meaning that the sasquatch was fourteen hands high.

It was a few years later that Grandfather had his second, but far more amazing contact with the Sasquatch. [A timber rattler had struck him in the leg while guiding men searching for gold. He had gone ahead of the group and was therefore alone when this occurred.] Grandfather killed the snake and started to come back down to a more comfortable spot, but soon found it difficult to go on, and as best as he can remember he became sick at his stomach and fainted. When he came around again, he thought he was dreaming, for three large Sasquatch about eight to ten feet tall surrounded him. He noted that they had made a small cut on the snakebite and had somehow removed some of the venom, and placed cool moss on the bite. Then one of the *matah kagmi* made a kind of grunting sound and the two lifted him up and took him down a trail that he did not know. Finally after some little descent down the mountainside, they placed him under a low brushy tree and left. Again Grandfather heard that mournful cry of the Sasquatch, *"Aagooooooouummmmm."*

After a while he began to feel better, and then took his old .44 caliber cap and ball pistol and began to fire some shots in the air. Finally the gold party found him. Grandfather said nothing about what happened concerning the Sasquatch. He was taken back to where the pack mules were tied, and then on to the nearest little town where he rested for a few days, and then returned to Tulelake. Grandfather told only his immediate family about this encounter, and after this would never take anyone for any amount of money to the Mount Shasta region. He would only say: *"matah kagmi* live! That Holy Place, I have friends there."

For many years after, in the still of the evening or sometimes late at night, he would still hear the sound he now knew, "Aagoooooouumm," the call of the Sasquatch. Grandfather went on to relate that the *matah kagmi* were not vicious, but were very shy, especially of the white man, and they generally only came out in the evenings and at night. They lived chiefly on roots they dug and berries, and only ate meat in the bitterest of cold weather. Their homes are in deep mountain side burroughs, unknown to man.

Original Source: "Encounters with the Matah Kagmi," *Many Smokes,* (National American Indian Magazine), Fourth Quarter, 1968. Modoc County, California.

Comments: This is one more most valuable story which I take for real. Hominology is never short of riddles and mysteries. The name *Matah Kagmi* is a great surprise! It's clearly a variant of the celebrated *Metoh-Kangmi* which in an incorrect translation became nown as the abominable snowman of the Himalayas, or yeti. Let me remind you that in 1921, Colonel Howard-Bury was on a reconnaissance expedition to Mt. Everest. They came unexpectedly across big footprints, which the Sherpa porters said belonged to a "creature of human form to which they gave the name *Metoh-Kangmi*" (Ivan Sanderson, *Abominable Snowmen,* p.10). How is it that the name turned up in North America in a Native language? Urgent help of linguists and ethnographers is required for us to investigate the mystery. Something interesting and surprising may be uncovered as a result.

"Return gifts" from homins to humans are fully in accord with barter trade between them. A sasquatch befriender (well-known to Dr. Henner Fahrenbach and some other bigfoot researchers) whom I call Lady Number One, because she prefers to remain anonymous, wrote me the following:

> I have left gifts [to sasquatches – D.B.] many times over the years and been given gifts in return. First was a pair of kittens, then a long haired domestic rabbit, a turtle, numerous "food" items and last a baby goat which was taken from the neighbor. Actually three goats were taken that particular morning before dawn, one doe and two kids. One of the kids was placed on my deck as the Big Guy passed my house. It woke me up circling the house on the deck bleating.

What is remarkable here is the fact that only the human animal, as far as I know, invented the custom of "giving gifts in return" (BRV, p. 382).

I wonder what tool was used to make a small cut on the snakebite—stone knife, steel knife or finger-nail?

Homins helping humans in distress is well-known. Back in the 1930s, a Russian family were gathering bast in the forest. Their daughter stayed separately with their horse. Decades later, being elderly, the daughter related to Maya Bykova what happened next:

> "I noticed a gadfly on the horse's hind leg and took a stick to drive the gadfly away, but the moment I touched the horse's leg with the stick, the animal involuntarily kicked and hit me. I fell to the ground. I remember hearing my brother yell loudly in panic, calling for mother and granny. At that moment I can recall the sensation of being lifted up and carried quickly away. Next I felt cool water running over my head. I opened my eyes: bending over me was a horrible human face. It was covered with hair, like the rest

of the body. I screamed. Back came granny's desperate cry. (...) Later grandma told me they'd found me, not on the clearing, but on the edge of a pond called Wolf's Hole. The creature had fetched handfuls of water and poured it onto my head, looking around all the time. When mother spotted him, she yelled and he immediately ran off into the bushes" (FRS p.179).

The Puyallup (page 202)

The Puyallup historically occupied western Washington. Today, their reservation is located near Tacoma. Their language is in the Salishan linguistic stock. The Puyallup have a close relationship with other tribes in the area and share traditional names for bigfoot. These includes *Steta'l* ("Spirit Spear") and *Tsiatko* (Wild Indians), shared with the Nisqually, and Seatco (Stick Indian), shared with the Yakama and Klickitat.

The Demons in Spirit Lake

The lake at the foot of the beautiful mountain *Loo-wit* was the home of many evil spirits. They were the spirits of people from different tribes, who had been cast out because of their wickedness. Banding themselves together, these demons called themselves *Seatco*, and gave themselves up to wrong-doing.

The *Seatco* were neither men nor animals. They could imitate the call of any bird, the sound of the wind in the trees, the cries of wild beasts. They could make these sounds seem to be near or seem to be far away. So they were often able to trick the Indians. A few times, Indians fought them. But whenever one of the *Seatco* was killed, the others took twelve lives from whatever band dared to fight against them.

In Spirit Lake, other Indians said, lived a demon so huge that its hand could stretch across the entire lake. If a fisherman dared to go out from shore, the demon's hand would reach out, seize his canoe, and drag fisherman and canoe to the bottom of the lake.

In the lake also was a strange fish with a head like a bear. One Indian had seen it, in the long-ago time. He had gone to the mountain with a friend. The demons who lived in the lake ate the friend, but he himself escaped, running in terror from the demons and from the fish with the head of a bear. After that, no Indian of his tribe would go near Spirit Lake.

In the snow on the mountaintop above the lake, other Indians used to say, a <u>race</u> of man-stealing giants lived. <u>At night</u> the <u>giants</u> would come to the lodges <u>when people were asleep</u>, put the people <u>under their skins</u>, and take them to the mountaintop <u>without waking them</u>. When the people awoke in the morning, they would be entirely lost, not knowing in what direction their home was.

Frequently the giants <u>came in the night</u> and <u>stole all the salmon</u>. If people were awake they knew the giants were near when they <u>smelled their strong, unpleasant odor</u>. Sometimes people would hear three <u>whistles</u>, and soon <u>stones would begin to hit their lodges</u>. Then they knew that the giants were coming again.

Comments: The name "Wild Indians" is indicative of the "demons" human status. That they were "neither men nor animals" I interpret as they were neither ordinary men nor ordinary animals. If these beings really "called themselves *Seatko*," this means they could speak and had a language. The term "demon" used in this story means that demonology is the correct and legitimate discipline, among others, in the study of the multidisciplinary subject of hominology. "Spirit Spear" is an interesting term, and along with other usages of the word "spirit" here indicates to me the "supernatural" (paranormal) abilities of these beings. "They were the spirits of people from different tribes, who had been cast out because of their wickedness" is a frequent mythological explanation of these beings' origin, …including the beliefs that demons arise from dead people who were not buried or buried the wrong way" (BRV, p. 28).

"A demon so huge that its hand could stretch across the entire lake" implies that a fisherman was not safe from an underwater attack by a demon anywhere in the lake. Richard Bernheimer wrote of: "the wild men and women who inhabit not the woods, but the water… The creature in question is an ogre who dispatches people by pulling them under the surface of the water. Significantly the pond is located in the midst of a wild wood..." (*Wild Men in the Middle Ages*, 1970, by Richard Bernheimer, pp. 39, 40).

We learn from John Green that: "Two of the reports involve the sasquatch either disappearing under water or appearing from it, and the Indians I spoke to said that the creatures could not only swim a long way under water but could do so at tremendous speed," (*Sasquatch: The Apes Among Us*, 1978, p. 430). Semi-aquatic hominoid creatures are part and parcel of Russian folklore and demonology. That's what Lady X calls "patterning": "As I've found in studying these individuals, as well as in this and related subjects, patterning is a most, if not **the** most, valuable learning tool."

All in all, this is one more story tightly packed with educational information.

The Demons in Spirit Lake (continued) (page 203)

A *Seatco* is in the form of an Indian but larger, quick and stealthy. He inhabits the dark recesses of the woods, where his campfires are often seen; he sleeps by day but sallies forth at dusk for "a night of it." He robs traps, breaks canoes, steals food and other portable property; he waylays the belated traveler, and it is said to kill all those whose bodies are found dead. To his wicked and malicious cunning is credited all the unfortunate and malicious acts which cannot otherwise be explained. He steals children and brings them up as slaves in his dark retreats; he is a constant menace to the disobedient child, and is an object of fear and terror to all.

Original Source: *Indian Legends of the Pacific Northwest*, by Ella Clark, 1953. Berkeley: University of California Press, pp. 63–64.

Comments: That "his campfires are often seen" in "the dark recesses of the woods" is most important and significant. It coincides with information received by John W. Burns from the Natives and rejected as legendary by John Green. That kidnapped humans are said to be used as slaves is also noteworthy.

The *Tsiatko* (page 203)

In my grandfather's time, his people captured a *tsiatko* boy and raised it. The child slept all day, then went out nights when everyone else was asleep. In the morning they would see where he had piled up wood or caught fish or brought in a deer. Finally, they told him he could go back to his people. He was gone many years and then came back once. He brought his *tsiatko* band with him and the Indians could hear them whistle all around. He said he came just for a visit to see them. Then he went away for good.

Original Source: *Indian Legends of the Pacific Northwest*, by Ella Clark, 1953. Berkeley: University of California Press.

Comments: One of the many cases of young homins captured and raised by humans. The boy's behavior, typical of sasquatch and different from human, is emphasized. Not surprisingly, being in captivity, he was able to learn to speak.

The *Tsiatko* and *Seatco*

A race of tall Indians, called "wild" or "stick" Indians, was said to wander through the forests. In general conversation they were referred to as *tsiatko* although another term, *steta'l*, from *ta'l*, spear, could also be applied to them.

The *tsiatko* lived by hunting and fishing. Their homes were hollowed out like the sleeping places of animals and could not be distinguished as human habitations. It was largely because of this lack of any houses or villages that they were characterized as "wild." They wandered freely through the wooded country, their activities being mainly confined to the hours of darkness. As has been said, they were abnormally tall, always well over six feet. Their language was a sort of a whistle and even when people could not see them they often heard this whistle in the distance. They had no canoes nor did they ever travel by water.

The giants played pranks on the village Indians, stealing the fish from their nets at night, going off with their half-cured supplies under cover of darkness, etc. Sometimes pranks on the persons of individual men, such as removing their clothes and tying their legs apart, were made possible by a sort of hypnotic helplessness engendered by the sound of the giants' whistle.

The giants were dangerous to men if the latter interfered with them or caused hurt to one of their members. Under these conditions their hatred was implacable and they always tracked the culprit down until they finally killed him with a shot from their bows. Occasionally also, they stole children or adolescents and carried them off to act as wives or as slaves. For this reason children were mortally afraid of going about alone at night and the *tsiatko* threat was used in child discipline.

During the summer camping trips when mat houses with loose sides were used for shelter, children always slept in the center surrounded by their elders for fear that the *tsiatko* would lift the mats and spirit them away. Men avoided conflicts with the giants and women retained the fear of them throughout their lives. Thus, one informant, a woman approaching seventy, broke her habit of rising before dawn and going to an outhouse at some distance from her home because she heard the whistle of a giant one morning.

Original Source: *Legends Beyond Psychology,* by Henry James Franzoni III and Kyle Mizokami.

Comments: Mention of "their bows" is amazing, seemingly out of synch, although a quiver has been mentioned earlier. I can't reject it out of hand, and can't accept it as certainty until more information is available.

The Spokane (page 205)

The Spokane remain residents of their traditional homelands in Washington State. Their language is part of the Salishan. The Spokane word for bigfoot is *Sc'wen'ey'ti,* meaning "Tall Burnt Hair."

Scweneyti

Scweneyti (*Chwah-knee-tee*) is a [direct translation] "tall, hairy, smells like burnt hair" [being].] He is about nine feet tall and possesses a very strong stench. He never hurts human, but does like to play tricks on people, like throwing rocks at them. He loves to tease horses and dogs. One day a family heard strange indistinguishable sounds coming from a draw far up in the mountains. The sounds, as of a man crying, drew nearer and echoed and resounded throughout the mountains until they were very close. The dogs were barking hard and suddenly were thrown against the flap of the tepee. The daughter, the braves and smartest, went outside and said, "Why are you terrifying us this way? We're already afraid! We know who you are. We know you are Cen. That this is your punishment for the sin you committed, and here on this Earth there will be no end to your wanderings, roaming about. Then here you are, here terrifying us, scaring us. It is God's will that you be this way. Go! Turn yourself around and walk away! Get away! Back away from us!"

Suddenly all was quiet. After a little while they heard *Scweneyti's* voice again, but further away. He had left them, no longer bothering them and terrifying them.

Original Source: *They Walked Among Us-Scweneyti and the Stick Indians of the Colvilles,* by Ed Fusch, 2002, pp. 11–13.

Comments: Very interesting religious belief and explanation of the hairy being's predicament. I wonder if it's indigenous or formed under the influence of Christianity. If the former, then detailed knowledge of the belief is desirable.

Fish and the *Scweneyti*

Fish was known to be a favorite food in *Scweneyti's* diet, especially when cooked. With their racks full, the Indians would cook the heads and parts they did not want and leave them out for *Scweneyti*, who would not disturb their fish or camp nor torment their horses or dogs.

Original Source: *They Walked Among Us-Scweneyti and the Stick Indians of the Colvilles,* by Ed Fusch, 2002, p. 15.

Scweneyti is Captured and Bound (206)

While camped at Keller, Washington during the salmon harvesting season, Grandmother, two of her sisters, and her brothers' wives found *Scweneyti* <u>sleeping along a creek</u>. These three sisters and two other women, knowing that <u>when *Scweneyti* sleeps, he sleeps very soundly (he sleeps during the day)</u>, drove stakes into the ground all around him, then laced their braided Indian ropes crossed all over him, tying him very securely to the stakes. As he began to awaken they all <u>sat on him</u>, hoping to keep him down. He appeared to pay no attention to them and <u>rose effortlessly, breaking the ropes</u>. The women fell off as he arose and walked away. They had to <u>destroy their clothes because of the stench from their contact with *Scweneyti*</u>.

Original Source: *They Walked Among Us-Scweneyti and the Stick Indians of the Colvilles,* by Ed Fusch, 2002, p. 20.

Comments: I guess the story is made up and sounds so impressive and plausible because of vivid and true-to-life details.

The Wenatchee (page 207)

The Wenatchee lived along the Wenatchee River in Washington. Their traditional language is in the Salishan family. The Wenatchee word for a bigfoot-like creature is *Choanito,* meaning "<u>night people</u>."

Choanito

In the fall of the year, October, a group of male members of the tribe were on a hunting trip near Wenatchee Lake. One of the men became <u>separated from the rest</u> of the party and <u>was captured</u> by *Choanito*. He was taken to a cave far up in the Rocky Mountains and <u>held captive by a family</u> of

Choanitos <u>throughout the winter until spring. The odor in the cave was terrible</u>. They would not take him out hunting with them but made him remain in camp near the cave with the women. They were like <u>a different tribe of Indians. In the spring they returned him to the place where they had captured him</u>. Upon returning to his camp he was immediately recognized by the children who couldn't believe that he was back as he had been gone for so long. They thought that he bad been killed. He said that he <u>had been well-treated by *Choanito*</u>.

Original Source: *They Walked Among Us—Scweneyti and the Stick Indians of the Colvilles*, by Ed Fusch, 2002, p. 37.

Comments: This is one more realistic abduction story, strengthening even more already available patterning.

Choanito and Camas Root

A woman had dug camas roots which they [Native women] placed on the roof of their home, located near Nesphelem Creek, where animals could not get at them. During the night she heard *Choanito* on the roof. In the morning the camas roots were gone and *Choanito* had put her puppy up on the roof. *Choanito* is still very active in the area. At night <u>lights can be seen moving along the base of a nearby mountain as a pack of them travel along</u>, and many have been reported on Keller Butte. <u>People are always warned to be out of the mountains before dark</u>.

Original Source: *They Walked Among Us—Scweneyti and the Stick Indians of the Colvilles*, by Ed Fusch, 2002, p. 38.

Comments: "Zakirov said that, though such encounters are very rare, Forest Service rules prohibit their employees from spending the night alone in the mountains for fear of these wildmen" (FRS p. 118).

The Yakama (page 208)

In prehistoric times, the Yakama inhabited parts of the Columbia Plateau (Idaho, Oregon, and Washington). They now occupy a reservation in south central Washington. In their traditional Penutian language, the Yakama words for a bigfoot-like creature are *Seat ka* and *Ste ye mah* ("Spirit Hidden by Woods"). Because they were also closely related to other tribes, they shared additional names with the Klickitat, *Qah lin me* and *Qui yihahs* (The Hairy brothers); the Puyallup, *Seatco* (Stick Indian), and the Shasta, *Tah tah kle' ah* (Owl Woman Monster).

Stick-shower

The *Ste-ye-hah' mah* (tribe) or Stick-shower are a mysterious and dangerous people whose general habitat is the lofty forest regions of the Cascade Mountains. They haunt the tangled timber-falls, which serve them as domiciles, or lodges. They are as large as the ordinary Indian; their language is to mimic notes of birds and animals. Nocturnal in habit, they sleep or remain in seclusion during the day and consequently are seen only on very rare occasions. Under the cover of darkness, they perform the acts which have fastened upon them the odious appellation 'stick-shower.' It is then that they thrust sticks through any opening of the tepee or hunter's lodge, or shower sticks upon the belated traveler. The Indian who is delayed or lost from the trail is very apt to receive their attention. He may hear a signal, perhaps a whistle, ahead of him. Should he follow the sound, it will be repeated for a time. Then he will hear it in the opposite direction, along the path he has just passed. If he turns back, it will only to detect the mysterious noises elsewhere, leading to utter confusion and bewilderment. When the traveler is crazed with dread, or overcome by exhaustion and sleep; it is then that the Stick-shower scores a victory. Regaining his head, or awakening from slumber, the wanderer is more than likely to find himself stripped of all clothing, perhaps bound and trussed with thongs. He is fortunate to escape with his life.

Original Source: *Ghost Voices—Yakima Indian Myths, Legends, Humor, and Hunting Stories*, by Donald M. Hines, 1992. Issaquah: Great Eagle Publishing, Inc., pp. 52–53.

Comments: This material is impressive and informative.

Wild Stick-showers (page 209)

The wild Stick-showers live in the mountains, in <u>lodges underground</u>. Doors to lodges are heavy, <u>snow and earth. You cannot find them</u>. They <u>have no fire in these lodges. But they dry meat, dry salmon by fire somewhere in the woods where they hide. They dress in bearskins tied up the front with strings. Head of bearskin covers head of Stick-shower, keeps off rain and snow</u>. That bearskin dress is warm, is dry and warm for coldest winter.

<u>The Stick-shower is tall, is slender. He is good runner. He has medicine which gives him swiftness and strength.</u> (Some Indians claim he has medicine that renders him <u>invisible</u>.) <u>They go long distance in one night</u>. Maybe they hunt over on the n-Che'-wana (Columbia River) near Dalles early in the night. Next morning, they are over here in Yakima country, all up Yakima River. Stick-showers are <u>good hunters. Nothing can get away from them; nothing can escape them</u>.

When you hunt on Goat Rocks, you have to watch. You have to watch close all the time. You are on a rock; maybe you cannot see around that rock, cannot see on either side. The Stick-shower pushes you off that rock. You fall down, fall far down to death. Some [Indians] get killed that way. <u>To hunt where Stick-shower is, four or five of us go together</u>. Three hunt, walking not far apart. One is here, one down below. One is higher up the mountain. We watch ahead, watch on each side. Fourth is behind. He watches back over the trail. Stick-shower might be following us. Must always watch for the bad Stick-shower.

Original Source: *Ghost Voices-Yakima Indian Myths, Legends, Humor, and Hunting Stories,* by Donald M. Hines, 1992. Issaquah: Great Eagle Publishing, Inc., p. 54.

Comments: "No fire in these lodges" but "fire somewhere in the woods where they hide"... a pattern seems to be forming. "They dress in bearskins tied up the front with strings"— fact or fiction?

Whistling *Ste-ye-hah' mah* (page 210)

An Indian, whose house stood by the side of a lagoon beyond which stretched a deep forest, lay on his bed at an open window one evening. He heard a whistling out in the timber. He answered it, supposing that it was someone lost. In turn, he was answered from the trees and at closer range. This was kept up for some time, the voice in the woods often taking the cadence of a bird song or other forest sounds. The Indian began to feel "queer" and "out of his head." Surmising that he was being "fooled with" by the Stick Indians, he closed the window and remained in the house.

Original Source: *Ghost Voices-Yakima Indian Myths, Legends, Humor, and Hunting Stories*, by Donald M. Hines, 1992. Issaquah: Great Eagle Publishing, Inc., p. 55.

Comments: "...taking the cadence of a bird song or other forest sounds"—one more example of sasquatch sound imitating ability. It's no exaggeration to say that chimpanzees and gorillas cannot speak because they cannot imitate sounds (other than those of their own kind), while humans can and therefore speak. Sound imitative ability was evolved before the origin of speech and must have provided its indispensable basis. Sound mimicking ability in sasquatch is often reported, so they are well equipped with this basis for speech.

The *Ste-ye-hah'*

It is the delight of the *Ste-ye-hah'* to carry away captive children who may become lost or separated from their people. Many snows ago two little ones, a brother and a sister, were missing from a hunter-village in the mountains. The parents and friends instituted a wide search and found their trail. Small footprints showed between the imprints of adult tracks, and at various places the children had left bits of their skin clothing along the path. It was readily seen that they had been carried off captive. But by whom? No hostile tribesmen were in that region. The alarmed and fearful people continued their quest and soon came upon undisputable proof that the dreaded *Ste-ye-hah'* had possession of the little ones. Recovery was hopeless, and at a point where the trail disappeared entirely, the pursuit was abandoned.

Long afterwards, perhaps twenty snows, the parents of the lost children were camped in the mountains gathering huckleberries. One night while sitting in their lodge, <u>a stick was thrust through a small crevice in the wall</u>. The old man immediately called out, "You need not come around here bothering me, *Ste-ye-hah'*! I know you! You took my two children, *Hom-chin-nah* and Whol-te-noo. We are all alone since you took our children. Go away!"

The *Ste-ye-hah'* withdrew from the side of the tepee. He was the lost boy. When he <u>could not remember his native tongue</u>, he recognized his own name spoken by the old Indian, his father. He lingered about the lodge, all night, fearing to enter. As daylight appeared, he went back to his <u>people</u> and <u>told</u> his sister what he had seen and heard, that their own parents were in the lone lodge at the berry patch. The next night he returned to the lodge, but did not enter nor let his presence be known. The third night he came again with his sister and entered the lodge. He made the old people to understand that they were their lost children, Hom-chin-nah and Whol-te-noo. It was the bow and arrows of the old man hanging on the lodge pole that had deterred him from entering the previous evenings. <u>The children came often to see their parents, bringing them salmon in abundance. There has never been any salmon in that part of the Cascades, but the *Ste-ye-hah' mah* [tribe] had this fish in quantity</u>.

The old people went away with their children, who had married and had families of their own. Later, when Indians visited this place, only the empty lodge was to be seen. The parents stayed with the *Ste-ye-hah' mah* for one snow, then returned to the berry patch and rejoined their tribe. Ever since that time, when any of the Indians are in the mountains and hear <u>the Chief</u> of the *Ste-ye-hah' mah* hooting <u>like an owl</u>, calling to his people, they know the <u>mysterious beings</u> are abroad, <u>bent on mischief</u>. They listen. Presently they hear a cry like <u>some bird, or the chattering of a chipmunk</u> near their lodge. It is then that the startled inmates call out, "You need not come bothering around here! <u>I am a relative of Hoom-chin-nah and Whol-te-noo! This invariably secures that particular lodge from further molestation by the mysterious *Ste-ye-hah'*. They will not knowingly annoy the relatives of the two children whom they once captured</u> and who resided with them so many years as <u>members of their tribe</u>.

Original Source: *Ghost Voices—Yakima Indian Myths, Legends, Humor, and Hunting Stories*, by Donald M. Hines, 1992. Issaquah: Great Eagle Publishing, Inc., pp. 57-59.

Comments: This intriguing story is probably legendary, but unlikely based on the realities of homin-human interactions. "... a stick was thrust through a small crevice in the wall"—a stick again... Sasquatch befrienders, Lady X and Lady Y, repeatedly found little sticks placed in different arrangements near their houses, as they believe, by sasquatch. The ladies think these stick arrangements are signs with certain undeciphered meanings. This reminded me of an episode I witnessed in the Caucasus in 1964 during my first expedition there. A young Kabardian named Pate, from the village of Sarmakovo, gave me an account of his two quite realistic and credible sightings, and then added that a friend of his was cohabiting with an almasty. Our conversation proceeded as follows:

"How come?"
"Yes, she visits him three times a year. He has four children by her
"Where are they?"
"They stay with her in the wild."
"How is he dating her?"
"By means of little sticks."(?)
"So she doesn't speak?"
"She can say one word in Kabardian."
"Which?"
"Give!"
(BRV, p. 17, 18)

Is this sheer fantasy? Maybe, but what do we do with the emerging little sticks pattern on both sides of the Atlantic?

The Capture (page 211)

Two *Ste-ye-hah'* captured a Yakama man and carried him <u>on their shoulders</u> to their home. One of the captors wanted to take him <u>to his sister</u>, but the other wanted to kill him. At last <u>the friendly *Ste-ye-hah'*</u> slipped the man away and told him to escape to his own land. He said to him, "Hurry away! There is a tall tree on the ridge where you will be overtaken by the darkness. <u>Sleep on the top branches of this tree. The *Ste-ye-hah'* cannot climb the tree after you</u>! The Indian did as instructed, and the pursuing *Ste-ye-hah'* came to the tree early after sundown and <u>were under it all night</u>. The Indian could hear his enemies constantly, but when the dawn came they left, and the man came from his perch and made the rest of his way home before another nightfall.

Original Source: *Ghost Voices—Yakima Indian Myths, Legends, Humor, and Hunting Stories,* by Donald M. Hines, 1992. Issaquah: Great Eagle Publishing, Inc., p. 60.

Comments: True or not, this story touches upon the important topic of the tree climbing ability of sasquatch. They are primates, and yet cannot climb trees, while even humans can? In Eurasia we have many reports of homins climbing trees. Slavic rusalkas (they are females) like to swing on tree branches and jump from trees growing on river banks into the water. Perhaps the question regarding sasquatch is clarified by the following provided by Janice Carter Coy:

> The bigfoot young were more apt to climb trees and play in them. I never saw Fox in a tree except on the lower branches of an old oak tree once. Sheba could climb one in a minute if she was after a bird. Yes, she would catch and eat birds and steal their eggs out of the nest if she could get to them. If the nest was too high for her to reach safely in the tree because of her weight, she would send one of the young

bigfoot up to get the eggs. They climb really well but use their hands and arms to hoist themselves up like people do. One time Blackie broke the eggs he went after before he got down out of the tree and Sheba scolded him for it, or at least that looked like what she was doing. She was shaking her fist at him" (*50 Years with Bigfoot* by Mary A. Green & Janice Carter Coy, 2002, Green and Coy Enterprises, Tennessee, p.146).

So it must have been the difference in weight between a human and the Stick-showers that saved the life of that Yakama man.

The Cherokee (page 225)

(...) As part of the Iroquoian language stock, the traditional Cherokee names for a bigfoot-like creature are *Kecleh-Kudleh* (hairy savage) and *Nun' Yunu' Wi* (stone man).

The Snake with the Big Feet
Long ago, in that far-off happy time when the world was new, and there were no white people at all, only Indians and animals, there was a snake that was different from other snakes.

Original Source: Unknown.

Comments: Just tells what made that far-off time so happy.

The Chickasaw (page 230)

Forcibly relocated to Oklahoma in the 1800's, the Chickasaw were once a large tribe spread throughout Mississippi. Their language is in the Muskogean family and is very closely related to the Choctaw. Their traditional word, Lofa, means "smelly, hairy being that **COULD SPEAK**." (My emphasis – D.B.).

Wiley and the Hairy Man

Wiley's mama knew all about things that were magic, like the Hairy Man in the forest.
"The Hairy Man got your daddy, and if you're not careful, Wiley, he'll get you too!" Wiley's mama often warned.
"I'll be careful," Wiley promised every time.
Wiley had never once so much as caught a sniff of the Hairy Man. All the same, he felt better if he had his two dogs with him when he went into the forest.
One day, Wiley was chopping wood when a pig ran squealing by and his dogs raced after it. No sooner had they disappeared among the trees than something huge and hairy with sharp, pointy teeth came lumbering toward Wiley. It was the Hairy Man! (...)
"You say you can do magic," continued Wiley. "So can you make things disappear, like all the rope in the neighborhood, for instance?"
"Of course," said the Hairy Man, scrunching up his eyes tightly, then opening them again. "There—it's done!"
"Oh good!" cried Wiley. "My dogs were tied up, but now they'll be free. Hoooo-eeeeee!"
"Yikes!" yelped the Hairy Man, fleeing into the forest.
Wiley's mama was very proud of her clever son and she was excited too. She knew that if you could trick a monster three times, he'd have to leave you alone forever. (...)
"That's the third time we've tricked you, Hairy Man!" He grinned. "So now you have to leave us alone forever. Hoooo-eeeeee!"
"Yikes!" cried the Hairy Man, and Wiley's two dogs chased him all the way back to the forest.

Original Source: Unknown.

Comments: At the UBC sasquatch conference in 1978 (Vancouver, BC), Dr. James R. Butler contributed a paper entitled, "Theoretical Importance of Higher Sensory Development Toward Avoidance Behaviour in Sasquatch Phenomenon." That's the language, or if you wish "jargon," of science. In people's plain parlance, "Higher Sensory Perceptions" are simply magic. The story is fiction inspired by what people call the Hairy Man's magic.

The Wildcat Clan (page 231)

This clan differs from other clans principally in what its members eat. They seldom go out in the daytime but roam about at night in search of food. They do not, however, try to steal. They are swift of foot and when an accident happens to them they depend on their swiftness to escape. They care very little about women, but when they want anything they generally get it. They think more of their feet than of any other parts of their bodies and their eyes are so keen that they can see anyone before he detects them. (...)

Original Source: *The 44th Annual Report of the Bureau of American Ethnology*, by John R. Swanton, 1926.

Comments: Here the striking fact is that this most unusual information was reported back in 1926 by a scientific institution, the Bureau of American Ethnology (!). I wonder if anyone asked the Bureau what kind of people on Earth could roam about at night in search of food?

The Choctaw (page 233)

Originally from Mississippi, Alabama, and Louisiana, the Choctaw were forcibly removed to Mississippi and Oklahoma in the 1800's. Their language is part of the Muskogean stock and very closely related to Chickasaw. Their traditional names for a bigfoot are *Kashehotapalo* (cannibal man), *Nalusa Falaya* (big giant) and *Shampe* (giant monster).

Shampe

All of the evil spirits of the Choctaws have followed them on their long journey from the western part of North America.

The witches, demons, and the monsters came with the Choctaw people. But the most horrible frightening of all these beasts is the hideous monster the Choctaws call the *Shampe*. A *Shampe* is a giant in the form of the ugliest Choctaw beast. He lives in the <u>deepest part of the woods</u>. So far in the forest that no Choctaw has even been able to find the location of his huge, dark cave. The *Shampe* cannot stand the brightness of the sun or the open air.

The <u>smell of blood</u> will attract him and he will follow the person who has been hunting and carrying a <u>wounded game</u>. *Shampes* do not have very good vision but have a keen sense of smell. They can <u>track any person or animal</u>.

The *Shampes* make a <u>whistling noise</u> as he stalks through the forest. His <u>scent is so terrible</u>, that many people have died from his odor. While he looks like a gigantic form of Choctaw, he smells like a skunk. <u>Some of them are really hairy like an ape while others are</u> **HAIRLESS**. (My emphasis – D.B.). The Choctaws won't live in an area where a *Shampe* will live or has been spotted. The Choctaws will often be caught or chased by a *Shampe*. <u>If someone were to drop a small game such as a rabbit or a squirrel, the Shampe stops to eat it and may be drawn off your trail by the blood of the small animal.</u>

Shampes <u>have followed the Choctaw people along their journey from the western United States</u>. They say that all *Shampe* <u>have returned to the west now</u>. But today, some Choctaws <u>still hear whistling sounds in the woods and catch a strong odor</u>. The Choctaws <u>still drop small animals when they think a</u> *Shampe* <u>is near</u>. You may never know that all *Shampes* have returned to the west.

Original Source: Unknown.

Comments: That some sasquatch are hairless is sensational. My supposition is that this is caused by interbreeding with *Homo sapiens*. Here's more information in support of the sensation:

> There were other bigfoot that appeared almost hairless. They are the same height as the other bigfoot. The females are birth-a-butts too and look made for

child bearing. These hairless ones are the same in height and weight as the ones with hair. Their necks appear to be shrunk in too, like they have no necks at all. (...) He looked like a bodybuilder would with the rib cage really tight looking and the muscles rippled over his stomach area. (...) He was here only a few weeks to a month or so and he never really showed up much until dusk anyway. There was only one time that he came out in daylight where I could see him clearly. (...) The hairy and hairless bigfoot are sort of odd looking in their own respect. I've only seen two in all the time I have had them around us close enough to see what they looked like. To me one looked to have a cave man type look. It was one of the strange males that showed up here once and it took Cheeco as a mate, I think, as that was when she went missing. (...) The other one was a young female that showed up with the clan once and stayed only for a few weeks. She looked like the others except she didn't have a lot of hair, just hair on her upper arms and across her shoulders and on her legs and privates and under her arms and on her head. (...) She looked almost exactly like the man in the picture of the Neanderthal Stan sent that time to us that is in the book he gave us that night at the motor lodge. (...) I always wondered if maybe she could have been a cross between a bigfoot and a human. (...) She was wild and Papaw didn't ever attempt to get any closer than maybe a few 100 yards away from her. (...) She hit me once with a rock in the center of the back, and once with a clod of dirt in the top of my head. (...) Sheba and Cheeco did not like her at all. They would take every type of opportunity to hit her and run her off from the group. They threw things at her.

Source: *50 Years with Bigfoot* by Mary A. Green & Janice Carter Coy, 2002. Tennessee: Green and Coy Enterprises, 2002, pp.135, 136).

Comments: Can you imagine a hoaxer, smart enough or crazy enough, to be selling news of hairless bigfoots? It's a lesson to all who have been fooling themselves regarding Janice Carter Coy's evidence.

Little-Man-With-Hair-All-Over (page 234)

Little-Man was hairier than a skunk. Hair grew out of his nose and nostrils. He had thick, matted hair between his buttocks. He was not particularly good-looking and he smelled as if he didn't wash often, but he was a merry fellow who laughed a lot, and he never had any trouble finding pretty girls to share his blanket. He was always on the move, eager to discover new things.

Little-Man-with-Hair-All-Over was small, but he succeeded in everything he did. He was tough in a fight, so they called for him whenever there was something dangerous to do. When a bear monster went on a rampage, ripping up lodges with his huge claws and eating the people inside, Little-Man-with-Hair-All-Over had no trouble killing it. For this his grateful people gave him a magic knife. (...)

Original Source: Unknown.

Comments: This is folktale evidence of the existence of dwarfish homins.

The Seminoles (page 246)

The Seminoles inhabited portions of Arkansas and Oklahoma in prehistoric times, moving into Florida when it was still ruled by the Spanish. The tribe maintains reservations throughout Florida. In their traditional Muskogean language, *Esti capcaki* means "tall man" and *Ssti capcaki* means "tall hairy man."

Ssti capcaki

Tall Hairy Man or *Ssti capcaki* resembles a human being but of immense stature, ten feet or more in height, and covered with <u>gray hair. He customarily carries a great wooden club made from a branch broken from a tree</u>. Tall Man is reported to have <u>a penetrating odor</u>, like the smell of a stagnant muddy pond.

Willie Lena's father encountered Tall Man once when Willie was very young:

> When Daddy saw it he told Mamma and said that it looked like he had <u>made his club from a limb of one of the trees on our place</u>. Mamma said, "If that is so, that tree he broke the limb from will soon be dead!" We all doubted this, but surely enough, the tree died. Where the branches had been there were big holes. It is in holes like this that Seminole women bury stillborn babies. I used to hear a baby crying at one of these trees near our house. There were little bones in there.

Original Source: *Oklahoma Seminoles-Medicines, Magic, and Religion,* by James Howard and Willie Lena, 1984. Norman: University of Oklahoma Press, p. 211–212.

Comments: In Russian folklore, the Leshy (woodman, wood goblin) is also said to carry a great wooden club.

The Lytton Girls Who Were Stolen By Giants (page 267)

Once some people were camped on the hills near Lytton, and among them were two girls who were fond of playing far away from the camp. Their father warned them against the giants, who infested the country. One day they rambled off, playing as usual, and two giants saw them. They put them under their arms, and ran off with them to their house on an island in a large river, a long distance away. They <u>treated them kindly</u>, and gave them <u>plenty of game to eat</u>. First they brought them grouse, rabbits, and other small game; but when they learned that the girls also ate deer, they brought to them plenty of deer, and the girls made much buckskin. <u>The giants were much amused when they saw how the girls cut up</u>

the deer, how they cooked the meat and dressed the skins. For four days the girls were almost overcome by the smell of the giants, but gradually they became used to it.

For four years they lived with the giants, who would carry them across the river to dig roots and gather berries which did not grow on the island. One summer the giants took them a long distance away, to a place where huckleberries were very plentiful. They knew the girls liked huckleberries very much. They left them to gather berries, and said they would go hunting and come back in a few days to take them home. The elder sister recognized the place as not many days' travel from their people's home, and they ran away.

When the giants returned for them, they found them gone, and followed their tracks. When the girls saw that they were about to be overtaken, they climbed into the top of a large spruce-tree, where they could not be seen. They tied themselves with their tump-lines. The giants, who had lost their tracks, thought they must be in the tree, and tried to discover them. They walked all around and looked up, but could not see them. They thought, "If they are there, we shall shake them out." They shook the tree many times, and pushed and pulled against it; but the tree did not break, and the girls did not fall down. Therefore the giants left.

After they had gone, the girls came down and ran on. The giants were looking all around for their tracks, when at last they came to a place where the girls had passed. They pursued them; and when the girls saw that they would be overtaken, they crawled, one from each end, into a large hollow log on a side-hill. They closed the openings with branches which they tied together with their tump-lines. The giants lost their tracks again, and thought they might be in the log. They pulled at the branches, but they did not move. They peered in through some small cracks, but could not see anything. They tried to roll the log down the hill, to shake out whatever might be inside, but it was too heavy. After a while they left. When they were gone, the girls ran on as before, and after a time reached a hunting camp of their own people in the mountains. During their flight they had lived on berries and fool-hens. Their moccasins were worn out, and their clothes torn. They told the people how the giants lived and acted. They were asked if the giants had any names besides *Tsawane'itEmux*, and they said they were called *Stsomu'lamux* and *TsekEtinu's*.

Original Source: *Legends Beyond Psychology*, by Henry James Franzoni III and Kyle Mizokami.

Comments: This is more folktale evidence of abduction of humans by homins and of happenings in captivity. Such material shows people's interest and curiosity regarding this phenomenon. The story is legendary, but based on what happened more than once in reality.

The Hare (page 271)

The Hare have been known by several different names, including Slavey and Slave. Today, as in prehistoric times, the Hare occupy the Northwest Territories of Canada, the Yukon, northern British Columbia and Alberta. In their traditional language of the Athabaskan family, a bigfoot-like animal is called a "bushman" or *Lariyi n* and *Naka*.

Lariyi n (page 271)

A *lariyi n* is a human-like being who roams around in the bush during the summer and steals women and children. They are considered to be foreign people who lost their way and became transformed into evil dwellers of the wilds.

Bushmen make house under the ground. They stay there all winter. In springtime they come out. They never make fire. They kill moose, and any animal. They might have guns, but usually they have knives, snares. I do not know if they have matches or not. They might smoke tobacco, maybe. They wear any kind of hide in winter. They are just men. There are not women in bushmen. They steal women but not children. They are in all sorts of ages—old ones and young ones. When there is no grub, they die and lie on the ground. *Ewe'n* [ghosts] might come out from the bushmen, too.

During the winter, they eat fresh meat. Even in winter, there is no fire. One or two people live together. But never three or more.

They whistle. [It is taboo for the Hare to whistle in the dark.] They do not have dogs. I do not think they start forest fires. I don't know how they would do with mosquitoes. They speak white man's language. All the white people who got lost in the [Indian] wars became bushmen. I have never seen a bushman. But my dad saw a bushman's track.

Original Source: *The Hare Indians and their World,* by Sue Hiroko Hara, 1980. Diamond Jenness Memorial Volume, National Museum of Canada, Canadian Ethnology Service Paper No. 63.

Comments: This is a contradictory tale indeed. Enough mention of familiar things pertaining to our subject, but mention of "guns" is quite baffling. Let's hope one day we shall learn what it really means.

The Nelchina (page 272)

There is not a lot of detail about the Nelchina available. Likely the term referred to a variety of tribes in Alaska who spoke languages within the Eskimaun family. Although the *Gilyuk* were known as fearsome black giants who ate people, their apparent cannibal nature did not translate well into their traditional name, which means "The-Big-Man-With-The-Little-Hat." Nevertheless, according to Murphy, Green and Steenburg (*Meet the Sasquatch*, 2004) this name came about because from a distance, the creature's pointed head (sagittal crest) made it appear as though it is wearing a little hat.

Gilyuk

Gilyuk is the shaggy cannibal giant sometimes called "The-Big-Man-With-The-Little-Hat." The Indians knew that *Gilyuk* was around because they had seen his sign, a birch sapling about four inches through that had been twisted into shreds as a man might twist a match stick.

Original Source: *Sasquatch: The Apes Among Us,* by John Green, 1978. Surrey, BC: Hancock House, p. 336.

Comments: Twisted thick tree branches and saplings, reported in homin habitats, is clear evidence that such things are done by hands—by hands that have tremendous power. Igor Burtsev saw and photographed such evidence on Janice Carter Coy's property. He also saw and photographed in a forest in Tennessee wooden constructions which have become known as "markers." Similar wooden constructions are also found in Russia and Australia, and there is enough reason to believe they are made by homins. Igor is now intensely investigating this phenomenon and has gathered many photos of homin markers. One thing is already clear, as pointed out by Michael Trachtengerts, we now have signs of the wildmen's presence in this or that area, not only on the evidence of their tracks but also of wooden markers. What's more, the latter are far more lasting, and therefore easier to find than tracks.

Before I come to overall conclusions, one more interesting thing the reader sees at the end of the book by Kathy Moskowitz Strain has to be mentioned. In its Appendix A—**Traditional Native American/First Nations' Names for Bigfoot**—a total of 142 such names are listed, of which 125 are translated into English. The meanings of 17 are unknown. Of the 125 known names, 45 mean or imply man (Man of the Woods, Wood Man, Hairy Man, Big Man, Tall Man, Wild Man, Cannibal Man, Stick Indian, Bushman, Big Elder Brother, Night People, etc.); 36 mean giant and also seem to imply giant **man**; 4 mean **devil** or **demon**; 4 mean **bigfoot**, and only 1 means **ape or monkey**.

Note that apes and monkeys (i.e., non-human primates) do not inhabit North America, so we can ask why Native Americans applied this name to bigfoot. I had a similar question when writing my book on folklore in Russia. The Chuvash, living in the Volga region, have two names for the homin said to be in that region: Arsuri ("half-man") and Upate ("ape or monkey"). A folklorist, writing about this, wondered why the Chuvash use that second name, because apes and monkeys do not inhabit Russia. My answer is that the Chuvash learned of the existence of apes and monkeys not so long ago, while they've always known their wild man. So either they applied one of their wild man's names to apes and monkeys or, vice versa—used the name for apes and monkeys to indicate their wild man because of certain **likeness** between these beings. I think this reasoning also applies to the name Ba'oosh ("ape or monkey") used by the Tsimshian, North American First Nations people. This seems plausible because their other name for bigfoot is *Gyaedem gilhaoli* ("Men of the woods"). In this connection, it is incorrect that the Malays call the big red ape of that region (orang utan) "man of the woods." The name of the big red ape in the Malay language is "mias." The Malay term "orang utan," used for a **real bipedal primate**, was wrongly applied to the big red ape "mias" by the Europeans in the 18th century, thus covering up a great error of science.

Boris Porshnev was the initiator of the Soviet 1958 scientific expedition to the Pamirs in search of the "snowman." When the expedition returned empty-handed, he wrote in his documentary story, *The Struggle for Troglodytes*: "We were clearly unprepared to question nature without first properly interviewing the people who have for generations lived in the lap of nature." Folklore is one of the main testimonies of generations of people living "in the lap of nature." Kathy Strain states in her book dedication: "This work is dedicated to the Native people of North America. These are your stories. Thank you for giving us a piece of your knowledge about a creature that you have always known."

Most stories are sufficiently old, just as folklore itself. So why do we now learn, or begin to learn, from folklore so "late in the day?" The reason is that there did not exist a science or discipline for the study of our subject. You do not apply to a botanist to learn about the existence and nature of electrons and protons, nor to a zoologist regarding the existence and nature of "black holes." You need a physicist and an astrophysicist for that. So factual information regarding our subject could not be extracted from folklore and demonology before the existence of hominology and hominologists.

The first obstacle to overcome for a student of hominology is to realize that the words **DEVIL, GOBLIN, BROWNIE** and the like one comes across in folklore and demonology, do not mean immaterial beings, mythological beings, or "mental constructions," as put by one ethnographer. Boris Porshnev was first to realize this. His opponents said to him: "Your snowman is nothing but a wood goblin"—they meant it was pure fantasy and mythology. "Yes," answered Porshnev, "only vice versa, a wood goblin is a snowman." The second obstacle is that, in truth, there is fantasy and mythology in folklore and demonology, along with things described accurately and realistically. There are stories and tales, called in Russian folklore and common people's parlance, "bylichka," which can be translated as "happening" or "what really happened," and there are "skazki," or fanciful "fairy tales." We have seen both kinds of tales in Kathy Strain's book. So we have to use common sense to tell them or their elements apart. Science is "organized common sense" (Thomas Huxley). There are also things and cases making it difficult or impossible to decide at once whether we are dealing with fact or fiction; but then hominology would not be a science if all matters were clear and uncomplicated. "Patterning," as pointed out by Lady X, helps us in such cases.

Lady X coexisted with, observed and studied a family of bigfoot on her wooded property for six years. Her first encounter with them was sudden and dramatic. This is what she wrote me in response to my inquiries:

> I'm not a specialist in Native American Studies or culture, and thus would never attempt to critique or judge their oral traditions. I'm not in a position to do so, and never would. But I would certainly turn to their oral traditions as a resource and tool, as I've done, and have found much application.
>
> Relatedly, your recent shared message was of particular interest as I reflected on my own initial journey—striving to acquire behavioral information.
>
> When I discovered I had not only a group of visitors, but a recurring group, and not knowing anything about them, the first and most immediate thing I needed to know was what to expect behaviorally to gauge my level of safety, or what, if anything, might precipitate aggression or lead to endangerment. I immediately went online and found tons on sighting and track reports, and other miscellaneous information, but virtually nothing regarding behavior. Annotated descriptions of books on the subject at the time sounded the same, and there was no one identified or found as having had any ongoing contact and experience for consultation. In short, I was on my own.
>
> My first thought: Who would or might be familiar with

these beings and behavioral aspects? From home (on sabbatical) I instantly speed-dialed the college reference librarian (who over the years I'd developed a close working relationship), and she kindly pulled for me (and even ordered off-campus) every book and reference she could find on Native American oral traditions.

Cannibalism... kidnapping young women and children... intercepting forest wanderers and travelers... intimidating and chasing off fishers and hunters... tricking and playing pranks... stealing fish and meat...

Behavioral themes gleaned from oral traditions often appeared to relate to territoriality and spatial organization, resource competition, habitat and resource protection, wildlife protection, food-resource procurement and maintenance, mobility, reproduction, and others.

I found this information most valuable, and it was this information, in part, that dispelled my concerns as seemingly little applied... (...) There would be nothing in my lifestyle, behavior or actions that should disturb or perturb them. In short, I sensed the situation would be fine, and we should be compatible. I also, in retrospect, was certain they had already been present for quite some time, and I'd never been harmed.

I should note I also used such preliminary information and profiles gleaned from Native Americans to help design methods of study and tailor approaches.

The point to be made: I turned to Native Americans—their oral traditions—to gain bearings and insights into behavioral aspects and considerations, and later used as reference for comparisons. What first struck me about oral traditions was that these beings certainly didn't sound like animals, or do things animals would do. They sounded quite humanlike in behavior and action—able to reason, outsmart, verbalize and more, and they were described and referenced in human terms. At the time it seemed fanciful, but over time, as close contact and experience accrued, I found it was a most accurate assessment.

I can't imagine who could explain and publicize the importance and usefulness of folklore for us better than Lady X with this message. As she states regarding oral traditions, "what first struck me was that these beings certainly didn't sound like animals..." To the question addressed by me to Kathy Strain, "What is your impression of the status the Native Americans and their folklore ascribe to Sasquatch —human or animal?" She replied: "Dmitri, I would say that most Native people feel that bigfoot is a form of human. (...) I think Native people view bigfoot as a relative—but the kind you don't really want to invite to Christmas dinner."

It was a long time ago, before writing this paper, that I opened Richard Bernheimer's book, *Wild Men In The Middle Ages*. Opening it on this occasion, I was struck by this sentence on page 5:

> Heinrich von Hesler, in the fourteenth century, explains in his *Apocalypse* that wild men are "Adam's children in form, face, and human intelligence, and are God's own handiwork."

Because of my initial preconception, borrowed from Boris Porshnev, I didn't earlier believe the fourteenth century author that "wild men" are "Adam's children in form, face, and human intelligence," that is humans on the whole. As I've stated already, the book that made me "betray" my hominology teacher, Professor Porshnev, and start to think that bigfoot can speak, and therefore by Porshnev's own criterion, must be classed as humans, not animals, was the book, *50 Years With Bigfoot*, that was published in 2002 and provided to me that same year. Despite various opinions, it's a great revelational book. Thus even more credit goes to the two non-professional authors, Mary Green and Janice Carter Coy, for making a groundbreaking contribution to hominology.

The two books, *50 Years With Bigfoot* and *Giants, Cannibals & Monsters: Bigfoot In Native Culture*, as demonstrated with a couple of examples above, are complementary and mutually supportive. All previous books on the subject, my first books in English included, were for the most part introductory, dealing mainly with the homins' existence and appearance, but indecisive on the most important issue of all— the beings' evolutionary status, the question of their being human or non-human primates. These two volumes can be referred to as our first textbooks in hominology, demonstrating to me beyond doubt that bigfoots, and similarly their hairy relatives in Eurasia and Australia, are **human primates.** Besides their linguistic ability, another impressive and indicative characteristic is the variety of their psychological types, the existence among them of "good guys" and "bad guys," just like among ordinary humans; some bigfoots being ready to kill and eat people, while others are known to help and save people. I first heard of that from Janice Carter Coy, and now learned the same from Kathy Strain.

So as far as I am concerned, the most crucial and prickly conclusion has been chewed and swallowed. And it's only the beginning. Digestion comes next. That is, reflection, comprehension, and figuring things out. Terminology is a headache as usual— terms being the tools of thought. **Homin, hominology, hominologist** are useful and usable. A homin is a living (non-fossil) non-sapien hominid (hominin). How closely are they related to *Homo sapiens*? As closely as wolves to coyotes or as wolves to domestic dogs, judging by fertile interbreeding between homins and humans.

A homin is a human being, but seemingly not a sapiens human being, and that is why we need a special name for him. Accordingly, there is humankind and hominkind. What's the difference? Not in appearance, not in morphology, but in mentality, in ecology, that is existentially? The crucial difference seems to be in the relation with the environment, i.e., with nature. *Homo sapiens* have built civilization by "conquering" and "enslaving" nature, while Homin-the-Wild Man has existed sort of melting into nature. No sapiens aborigines have such close and intimate links with nature as homins do. So the best qualifier for them is nature. They are nature people, and each of them is a nature person. Like it or not, in terms of "otherness," each of us is either a "tech person" or "hi-tech person."

Now the great philosophic and sociological question is how on Earth these nature persons have managed to evolve and maintain for ages human intelligence under conditions of wilderness. The answer will probably be long in coming, but my hunch is that their "magic" powers have something to do with it.

It's really time for me to wind up, and before saying goodbye let me remark that there is still a source in Native American culture that has not been tapped. I mean Native sayings and proverbs. The reader will recall my discussion on non-Native sayings and proverbs provided on pages 29 and 30 *(originally a part of this paper as well)*.

CHAPTER 3
Hominology in the Balkan Peninsula

(Originally published on the Bigfoot Encounters website March 2010; currently on the Sasquatch Canada website.)

Chris Murphy received the following message from the late researcher Lloyd Pye and forwarded it to me:

> Recently a fan from Serbia named Goran Z. Milic sent me an incredible piece of new information in this field. It is a direct quote from a very old Serbian school textbook designed for the 3rd and 4th grades of primary schools across Serbia. It was printed in 1898, 53 years before Eric Shipton photographed the famous string of yeti tracks across the Menlung Glacier in the Himalayas in 1951, which put such creatures into the modern consciousness. Until that point they were known only by the people who lived in the specific regions where hominoids lived, as was the case in Serbia. Here is the quote using Goran's translation:
>
> "There are also people that don't live in houses, nor do they know what houses are. They live in holes and caves, usually hidden. These people walk naked, and their bodies are covered with long hair. They don't know anything about clothing. They eat fruits, or raw animal meat that they hunt. They care about food only when they are hungry. If they have any extra food, they don't save it for some other time, because they don't seem to be aware that they will be hungry again."
>
> Relative to that last line, I would suggest they don't worry because the deep forests and jungles where they live are a literal cornucopia of food if you know how to recognize it and utilize it. Why eat leftovers when you don't have to? But in general, this is an astonishingly accurate description of how all four groups of hominoids live around the world, including the "pygmy" types that live

in the band of jungles around the equator, best exemplified recently by the discovery of the "Hobbits" of Flores Island in Indonesia.

I in turn shared this with researcher John Morley who came back with:

> Dmitri, this is most fascinating! I am curious as to the source of the material by the author of the book? Were first hand accounts involved, or did the words result from work or writings of someone else? How is it that in 1898 this knowledge was known to these people, and who were the ones who knew it?

I wish I could answer these questions. Yes, it's fascinating and marvelous that in 1898 the 3rd and 4th graders of primary schools were taught knowledge which is unknown to most scientists today, in the year 2010.

I then began thinking about the country—Serbia—where this knowledge was propounded, and more generally about the Balkan Peninsula, in the South-East of Europe, of which Serbia is a part. Another country of the Balkan Peninsula is Greece, and the specific role of Greece in hominology is in providing us with a lot of most valuable historic material. After I was introduced to hominology and underwent a process of demythologizing regarding a number of mythical names and creatures, the names of Pan, Silenus, satyr, and nymph were among them. Those were names that ancient Greeks gave to their homins whom they worshiped as gods and semi-gods—that is, lords of nature. I recall that when I acquired this knowledge, I paid several visits to the Pushkin Fine Arts Museum in Moscow, one of the world's richest, and looked, with quite different eyes and keenest interest, at ancient sculptures and pictures of satyrs; feeling pity for other museum visitors who didn't know what I knew.

We know of satyrs not only from sculptors and painters, but also from the ancient author Plutarch who wrote of a satyr that was captured while it was asleep by the soldiers of the Roman general Sulla in 86 BC, in the area of what is now Albania (another country of the Balkan Peninsula, neighboring upon Serbia). Plutarch describes the captive: "exactly such as statuaries and painters represent to us," and continues: "He was brought to Sulla, and interrogated in many languages who he was; but he uttered nothing intelligible; his accent being harsh and inarticulate, something between the neighing of a horse and the bleating of a goat. Sulla was shocked with his appearance and ordered him to be taken out of his presence."

The instant repulsive reaction of the general at the sight of a non-sapiens hominid gives a hint regarding the future torturous path of hominology toward scientific acceptance and recognition. And as to the satyr's harsh and inarticulate accent, there was a time when I believed that was proof of his lack of linguistic ability. Today, aware of Janice Carter Coy's description of speaking bigfoots, I am not sure of that at all. According to her, a speaking bigfoot would not be immediately recognized by a stranger as using language. She wrote:

> They do talk deep in their chest and it does not sound like human speech... Most of the time when they speak any given word it comes out in a long drawn out word or that of several words run together. This is especially true if they are yelling at you from off away from you somewhere. If they are near, they talk in voices that sound like a chatter, and it is rapid and deep and gurgled. Anyway, what I would call their own unique language would not be considered a language at all but an inarticulate one. It consists of screams, howls, chirps, whistles, clicks and clucks of the tongue and throat.

Scott Nelson, who supports Janice's claim of bigfoots being able to speak in some sort of language, wrote:

> ...though the creatures are speaking in language by the human definition of it, they are also making sounds that I don't believe humans can make... Many utterances are in frequencies that are beyond the ability of humans. (...) these voices are very different than any human voices that I have ever studied.

So the satyr's inability to explain to his captors who he was does not necessarily mean he was as dumb as a fish.

What became of satyrs and their ilk in the next historic epoch, namely the Middle Ages? With the advent of Christianity, all heathen cults and beliefs were condemned by the Church, and heathen gods and semi-gods were declared to be demons. Christianity won the day initially in the cities, while in villages and the countryside heathen cults went on to be observed clandestinely throughout later history. That is a theological and demonological aspect of the matter. Another aspect is that of natural history, and in this respect non-sapiens hominids became known as "wild men." Historical records of them in medieval Europe are very abundant. A lot of them are presented in Richard Bernheimer's famous book, *Wild Men in the Middle Ages*, as I have provided on page 17.

If after that you wonder why hominology is still a Cinderella in the world of science, you should read the first sentence in Bernheimer's Preface to his book where he explains that he tells "the case of *the imaginary figure* to which this book is devoted..." (My italics – D.B.). The author thus informs the reader in advance that the wild men of the Middle Ages are imaginary figures and his monumental, richly illustrated, volume of 224 pages deals with nothing but figments of the mind.

So let's return to the Balkan Peninsula and follow the fate of "imaginary figures" in modern times. In 1984, Western Publishers, Calgary, published the book, *The Sasquatch and other Unknown Hominoids* (1984), compiled and edited by Dr. Vladimir Markotic´and Dr. Grover Krantz. It contains contributions by authors from different countries, including one from Croatia, bordering on Serbia, both being parts of former Yugoslavia in the Balkan Peninsula. The author presents and discusses Croatian folk tales on the subject of our interest. As folklore is a pillar of hominology, I re-typed the article (with some omissions) for benefit of those who do not have the book, underlined words and sentences of special interest and inserted notes as to my comments.

Creatures from the Bilogora in Northern Croatia
by
Zvonko Lovrencevic

A considerable number of old stories and legends from the region of Bjelovar in Northern Croatia tell about creatures that possess <u>supernatural abilities</u> or <u>differ physically from</u>

people. (...) The present writer has collected stories about them for many years. Leaving aside a number of these stories, for the present we shall pay attention to the stories about Vedi and woodland Lasses. The stories about these creatures are so realistic and vivid that their very existence in the near past should not be completely doubted.

Vedi (Note 1)

Once a vast range of woodland covered the territory between the Bilogora mountains and Kalinovac and Ferdinandovac near Drava River. Various kinds of wild animals, as well as some strange man-like creatures lived there. People from the region called them Vedi. **(Note 2)**
Vedi were very tall male creatures, higher than houses. **(Note 3)** Their bodies were very much like man's, except for being covered with hair. When they talked or sang, one could hear them from afar. They were very strong—they could pluck out whole trees by their roots, and carry heavy loads. They could even raise a storm with their breath—they had very strange chests **(Note 4)**. It seems that Vedi wore clothes but very poor ones if one judges according to the saying, "Ragged as a Ved!" which is heard rather often in the region even today. **(Note 5)**

Note 1: In the Slav languages, Russian included, (Croatia is a Slav country), "i" indicates the plural of words, like "s" in English. So Vedi is the plural of Ved.
Note 2: There is a note at the end of the article which I have inserted here: "Ved may be related to 'wood' in English and similar words in other Germanic, and in the Celtic language, and the root in the proto-Indoeuropean languages is believed to be *widhu* (Klein 1971:804, 831). (...) Thus Ved could have originally meant "forest man."
Note 3: Some of the inhabitants of the region think that Vedi were not all that tall.
Note 4: Perhaps, barrel-like?
Note 5: Parallels elsewhere: "Demons can wear clothes, given by humans or stolen from them. The clothes are usually old, tattered, and worn inside-out"—BRV p. 48. According to Janice Carter Coy, bigfoot Fox sported a T-shirt of the largest possible size, bought for him by her grandfather, Robert Carter Sr. The bigfoot wore it until it turned into rags, then he continued to wear its collar round his neck.

<u>They were living in groups</u>, and in their own settlements <u>in dark woods</u>. What those settlements looked like and how big their houses were, <u>we do not know</u>, but they must have been very large because in certain clearings large bricks were found. **(Note 6)** <u>There were good and evil Vedi</u>. Evil or wicked Vedi did not associate with people. They remained in the woods, and that is why they were also called "woodland Vedi." They did not kill people, but if they found a young man in the woods, <u>they would take him</u> to their houses, and keep him as a slave there for some time. They would very often maltreat him, but then they would let him go home starved and exhausted.

<u>Good Vedi used to come near people and even helped them</u>. There was almost no household without a Ved. He was so loyal to his master that he would do all kinds of mischief to the neighboring household and their Ved—such as scatter cattle in the pasture, move a beam of the roof of a stable or a house so that the roof would incline or even collapse. The following is one of these stories:

There was a man who had his Ved and this Ved always helped him around the house. He had a cherry tree in the garden which bore him a lot of fruit. Once, up there on its top branches, he saw the most beautiful and the reddest cherries of them all, but was not able to reach them. The man called his Ved and told him to bend the top down so he could pick them. <u>The Ved bent the top branches down</u>, and his master started to pick and eat the cherries. However, as the man was eating, the neighboring <u>Ved started to poke fun</u> at his Ved, who suddenly <u>jumped over the fence</u> and started after him. The man was holding fast to the top of the tree, and as the Ved let it go, the tree sprang back to its upright position, dragged the man up, and up it threw him, up into the air, so he never returned back to the Earth. **(Note 7 – Next Page)**

Note 6: Petar Zrinski (…) correctly assumes that the bricks in question were from Roman ruins, i.e., Roman buildings that were abundant along the road leading from Petovio or Petavione to Mursa. (So these bricks had no connection with Vedi – D.B.)

A Little Digression: At this point I wish to digress a little on the subject of homins interacting with humans. There are stories in Russian folklore of homins helping hunters, fishermen, and shepherds. Also tales of homins (connected with neighboring farmhouses) quarreling and fighting among themselves. Peasants said that when domovoys (Russian brownies) are fighting, it is necessary to cry: "Hey, our own, beat that stranger!"

In her book, Janice Carter Coy presents a vivid picture of fighting bigfoots, as well as an incident when Fox saved his good friend Robert Carter Sr. from being gored by an angry bull. The whole of the Carter Farm story is an illustration of the homin-human interactions mentioned in this article by the Croatian author; or rather vice versa, this article is an illustration of the Carter Farm story. Similar cases regarding North America are to be found in my work, *Learning from Folklore*. As for Australia, researchers Tony Healy and Paul Cropper tell us of a Yowie's "working relationship" with a group of Aborigines:

> Even more remarkable were their accounts of an apparently unique, amicable relationship that once existed between the Goodjingburra and one particular Hairy Man. When this relationship began is unclear, but it supposedly continued even after white people began to settle on the peninsula in the mid-1900s. (...) Up until the early 1900s the Goodjingburra would travel up towards the mountains at different times of the year, "but instead of carting all their gear, fishing spears etc., they'd leave it here in one of the caves and the Hairy Man used to look after the stuff. I'm not sure if it could actually talk, but it could communicate" (*The Yowie*, 2006, pp.78, 79).

Note 7: This humorous and fabulous tale, cooked up of factual ingredients, gives an idea how tall, strong and athletic Vedi were. We have numerous stories from folklore and witnesses in Europe, America, and Australia of homins associating with humans in a friendly way. "Brownie, in the folklore of Scotland, was a goblin of the most obliging kind. He was never seen, but was only known by the good deeds which he did. He usually attached himself to some farmhouse in the country, and he was only noted by the voluntary labor which he performed during the night" (*The Everyman's Encyclopedia*, 1913).

(Creatures from the Bilogora...continued)
Vedi used to be good and useful to their masters, but evil and harmful to the others. When something evil happened, a host would find a neighboring Ved guilty although the harm could have been done by another man or perhaps bad weather. If a storm, flood, or some other disaster were expected, people would pray even in church, "Oh Lord, let our Vedi help us!" or "Beloved Lord, let our Vedi help us, and protect us from their Vedi!" **(Note 8)** Very soon after these kind of prayers, Vedi would come and help. It is still not explained whether they came flying or running. It must have been that they only ran, if one judges according to the saying "He came running as quickly as a Ved!" **(Note 9)**

Although it is thought by the people of the mentioned region that Vedi were males, some of the reporting people, while talking about Vedi, used adjectives and other expressions that indicate the possibility of there being also female Vedi. (...)

However, it is not impossible that some of the characteristics of Vedi were spread over to "woodland lasses." Woodland lasses were said to be living in those woods at the same time as Vedi. (...)

Vedi had their names too. Some of them had the same names as the fields and meadows, as for example Miklici, Pastacki, Busica, and such. **(Note 10)**

It is even said that some of the families were named after their Vedi. The case in question is the family name Patacki, which is rather frequent in the region. **(Note 11)**

In spite of their height and great strength, it is believed that Vedi were not immortal like fairies, gnomes and other

Note 8: This is a remarkable example of remnants of heathen beliefs. Note the words "supernatural abilities" in the beginning of this article. It was not for nothing that wild men were once believed to be gods and semi-gods.
Note 9: Let us note again that homin characteristics are "fossilized" in popular sayings and proverbs, which are a good source of evidence.
Note 10: I am still greatly fascinated by Janice Carter Coy saying her bigfoots have their own and rather poetic names.
Note 11: Bernheimer notes a similar phenomenon regarding medieval wild men.

creatures from the woods. <u>They died, but how and why we do not know. People came to this conclusion because they found skeletons of creatures with leg bones longer than 70 to 80 cm.</u> **(Note 12)** While ploughing, a man even found a "cemetery of Vedi." Dead bodies were buried in a very strange way—two large skeletons facing each other were found <u>in big hollow tree trunks</u>. **(Note 13)**

The last information about Vedi date back to the middle 19th century. Gradually, they started to leave people, fields and meadows, and retreated into the woods. At the end of the century Vedi were scarcely mentioned. They would appear here and there, someone would talk to his ex-Ved, but this grew rare. Something even more strange happened. Vedi became invisible during the day or remained visible only to certain people, under what conditions it is not known. **(Note 14)** (...)

According to some old peoples' stories, a man called, Solo lived with Vedi until his last days. He went to the woods to meet them, and they also visited him and helped him with his work, but nobody saw them. Vedi gradually became visible only at night. The number of people who could see them grew less, and even their names faded away in human memory, until they became legendary creatures.

Once in Palanci, there lived a man who always kept company with Vedi. Whatever they would do, he would do also. If they went somewhere to fight, he went with them. They would fight during the night. But before they left, they would go to

Note 12: In the Caucasus, a hundred years ago, village boys hiked in the mountains and found "huge bones" in a cave. When they returned to the village and told the elders about the find, the following dialogue took place between the boys and the elders: "And the upper leg bone was how long?" —Teddua marked two feet on the floor. —"And the lower one?" —He added another two. — "I think," Miriani said, "you found the place where the Giants came home to die."—"Were they men?"— "Yes," Miriani said, "different from us, but men." (FRS, pp.44, 45). North Americans should not think that their bigfoots are the only giants on Earth!

Note 13: Was this perhaps the origin of wooden coffins?

Note 14: I find information in this paragraph more subjective than accurate.

one of the village stables to take the best horses and ride away. They would fight and ride all night long, and in the morning they would return the horses to the stable from which they took them. However, those horses were so very tired and all covered with foam in the morning, so that they were of no use in the fields. People cursed the man and his Vedi but nothing helped. He went around with Vedi until his death. **(Note 15)**

During the First World War, the stories about Vedi were very much alive. There were a lot of lonely women whose husbands were fighting in the war, and who were running their farms by themselves. People gathered to dance, eat and drink in their houses during the night. (...) When a girl or a woman would like to leave the company late at night, one of the group would immediately start telling stories about Vedi who roam around and scare people to death. **(Note 16)** After such a story no one would dare leave the party before dawn.

After the First World War, about 1920, mothers used to scare their restless children "Beware, if you are not good I'll give you to Vedi!" **(Note 17)**

After the Second World War the stories about Vedi almost died out. Only an old man or woman would remember them.

Woodland Lasses

Woodland Lasses (*sumske dekle*) lived in the Bilogora region also, but they differed somewhat from Vedi. There are still a lot of stories about woodland Lasses in Bilogora, especially in the area of Novigrad, St. Ana and Ferninandovac. Among stories that are being told, here are the most interesting ones:

Note 15: Cases of homins riding horses, and with the effect mentioned above, are recorded in folklore and by witnesses in Russia, but not in America yet. Braiding of horse manes is reported both in Russia and America.
Note 16: There are familiar stories and "bogies" around the world.
Note 17: Also ubiquitous adults' ploy of very old origin, not "about 1920." Some of the author's conclusions seem hasty and unjustified.

In the woods of Bilogora, Kalinovac, Kalinovacki konaci and Virje there lived woodland lasses. In some villages of the area people used to call them "wild lasses."
<u>They were somewhat lower than humans</u> (? – D.B.), <u>but in every other way they were very much like women. Their bodies were covered with long, thick hair. Their faces were not very hairy, so one could clearly distinguish eyes, mouth and nose. Their hands</u> (arms? – D.B.) <u>were longer and stronger than those of humans, they could run very quickly, and in general, they were very strong</u>.
They did not wear clothes. They were never heard to talk to people or to each other, so <u>nothing is known about their language</u>. What is known is that they were heard shrieking in the woods. When they were hungry they squeaked and screamed.
They would leave the woods with the first fogs on the meadows, and <u>approach abandoned shepherds' fires to warm themselves</u>. **(Note 18)**
They avoided people because <u>they were very timid, but some instances were known when they visited houses. It is told that they even entered courtyards when very hungry. Housewives used to leave them some food in a corner of the yard. "They were grateful for every morsel, and showed their thanks to the housewife usually by sweeping the courtyard, working in the stable, bringing fire-wood, and helping in other ways</u>. (...)

— — 0 — —

It was a very cold winter. My brother and I were sleeping in the stable. Once, during the night, I woke up and felt that someone was between us. I touched it, and as it was hairy and warm as a human, I thought at first it was our dog. However, I touched it again and felt that it was a female being. I got frightened, and as I wanted [went] to wake up my brother, the hairy woman felt I was awake and started to draw herself from between us. I called my brother, he woke up but the woman was already at the door. We started towards the door, but she opened it and ran to the courtyard. We started after her, but <u>she was quicker</u>, she <u>ran over the fence</u> to the garden, and on <u>toward the woods</u>. We did not want to chase her

Note 18: Same thing in Russian folklore.

because the snow was very deep, but we saw her very clearly under the moonlight. It was a woodland Lass. Later on, we told all about it to my father, and he said that it was nothing to wonder at because <u>woodland lasses often came to warm themselves beside the boys during cold winters</u>. **(Note 19)**

— — 0 — —

In those times, all married men used to sleep in the house, while the grown sons slept on small beds in warm stables during the winters, and in barns during the summers. Thus they were free to go to the village at night. During their roamings they <u>usually met woodland lasses</u>, but they always ran away from them, or <u>at least nothing certain is known about their relations</u>. However, <u>women used to whisper among themselves, but what they knew nobody ever found out</u>. **(Note 20)** Those older ones [women] would joke with young men, "Beware, if I give you to the woodland lass, it will be hard for you!"

If a young man doubted the existence of woodland Lasses, women would swear, "God willing, I did not move from this place if there were no woodland Lasses. <u>I saw John with one of them</u>."

Woodland Lasses were not ill-natured or revengeful, and they never did any harm to man. People sometimes used to maltreat them, but they never revenged themselves.

— — 0 — —

Far, far away in the Blue Woods—

There lived a man who had a very well fenced pig shelter so that no one would take away or steal his pigs. He also kept a beehive in the yard from which he gathered honey in

---

Note 19: I take this story seriously, and what a wonderful story it is! One more striking example of how unique the beings we study are. Can you imagine any other primate in the wild, say chimpanzee or gorilla, behaving like that woodland lass, warming herself beside the boys?

Note 20: I witnessed women whispering among themselves on the same subject during my expeditions in the Caucasus, and one young man told me that his friend had children by a "woodland lass," called "almasty" there. The children allegedly lived with her in the forest. Chapter 5 in Richard Bernheimer's book has the heading, "The Erotic Connotations."

the autumn. Once he noticed that someone had entered his fold. It was after the rains, and he saw <u>a human footprint in the mud</u>. There was no trace of mischief, all the pigs were there, and no one had even touched his beehive. He was amazed. Another time when he was driving his pigs home from the pasture at sunset, he noticed that something was hidden under the beehive. It resembled a man but wasn't one, it resembled a bear, but wasn't one; one would say <u>it was a man covered with a thick hairy blanket</u>. The man did not get scared; he took a stick and started to beat as heavily as he could. The creature screamed, covered its eyes with its hands, and the man realized that <u>it was not a human voice, but the voice of a woodland Lass</u>. He immediately stopped beating it, but just poked it a little with the stick so it would leave more quickly. After that, it used to come back again. The man never drove it away because it was hiding in his pig shelter for protection from other wild animals, but it never did either him or his belongings any harm.

— — 0 — —

Nothing is known about their origin. How are they born and do they ever die? Do they have children? These have remained secrets which we will never find out. Those generations who knew the answers disappeared at the end of the 19th century. **(Note 21)**

Source: Vladimir Markotic' and Grover Krantz, Editors, 1984. *The Sasquatch and other Unknown Hominoids.* Calgary, Western Publishers, pp. 266–272).

Well, "How is it that in 1898 this knowledge was known...?" The answer to John Morley's question is no longer inscrutable. The information was all there for those who wanted to use their heads. Why then was it "unknown" to science? I answered long ago: because there was no science to know it. It's not a witticism, but

Note 21: What I have learned from hominology is that in all ages, hairy wild men, whatever their names and habitat, have always been cryptic denizens of nature, always subjects of mythology, folklore and hearsay among countryside people; "the erotic connotations" being one of the reasons for that.

mere statement of fact. As also noted earlier, the situation was analogous to the history of meteoritics. Stones falling from heaven have always been known to people, but not to science because there was no science of the phenomenon. Scientists regarded all reports of stones that fell from heaven to be mythology and folk tales, which in truth they partly were. To tell meteorites from usual stones, witnesses and oaths were not enough: a self-consistent scientific discipline was needed for that. As soon as it was built, the science of meteoritics has demythologized popular and religious beliefs connected with the phenomenon.

Our situation is similar. From the information available to date, the hairy bipedal primates we study appear to be on the human side of the divide between human and non-human primates, and thus no longer subjects of zoology, but anthropology. At first sight, this makes our task easier, for there is no need to get as proof a type specimen, demanded by zoologists/biologists. But in reality the task becomes harder than we ever imagined. By all indicationss, we are dealing with non-sapiens hominids, which is neither surprising nor unnatural. It would be surprising and unnatural if all hominids except ourselves had died out. But then all modern encyclopedias and anthropology textbooks solemnly tell us as indisputable fact that all pre-sapiens hominids died out in prehistoric times, and the only hominid survivor on Earth is our own glorious species. If this stumbling block were not enough, there is another, which is surprising and mind-boggling even to the hominologist. It's the fact that these non-sapiens hominids inhabit, side by side with modern humans, the six continents and have been reported in every state of the U.S.! No wonder such "fantasies" are viewed by academics as modern mythology.

Are we strong enough at present to refute and "demythologize" academic beliefs? We are up against a coalition of mighty forces: the cover-up by the U.S. government, strong corporate business and clerical interests, deep preconceptions and myopia of scientific orthodoxy. Against this giant force we are smaller than Lilliputians in front of Gulliver. No amount of evidence can make us victorious against a coalition of formidable adversaries. Truth is our only tool and weapon. So our only viable strategy can be in building up hominology as a self-consistent scientific discipline, with its own organization, staff and journal. It is with the authority of a scientific discipline that the world-shaking truth of our research can be made believable. Helped by science, the truth shall topple untruth and appear for good in textbooks, including those for primary schools.

Finally, the latest from the Balkan Peninsula. Kosovo is a war-ridden part of Serbia, which has declared its independence, unrecognized however by Serbia and some other countries. In 2005, researcher Will Duncan received a surprising message about a homin sighting in Kosovo, which prompted my June 15, 2005, email to some people. I repeat it in part here:

Some Thoughts triggered off by the Kosovo Sighting

Robert Kinion, of the U.S., was stationed at the FOB (Forward Observation Base) in Kosovo as a civilian contractor, and while driving to Camp Monteith to do a distribution run and pick up laundry and such, in the morning of February 15, 2005, sighted a hairy biped, that he calls "hominid," coming down a fairly steep slope and then crossing the road in front of the vehicle. For details visit http://www.cryptozoology.com website.

The witness was, in his words, "afraid of ridicule" from his work mates and feared for his security clearance, "so kept quiet about it," but after talking to his wife felt that he "can share it with like-minded peeps."

We have a lot of old-time "wild man" information from Western Europe, but only few, no more than half a dozen, present-day sighting reports west of Russia. Now it took an American (!) to add one more. He was observing and reporting possible unlawful actions by humans in the area and ended with observing and reporting the "unlawful" presence of a relict hominid. What a marvel! Being familiar with the bigfoot problem, the witness knew how highly such information is valued by those who are after it, and he knew how to bring it to the attention of interested researchers. Imagine scores of daily sightings around the world by those who lack such knowledge, or even have it, but do not report what they saw for fear of ridicule or something worse.

CHAPTER 4

Wheatcroft's Orang Pendek Evidence – Thoughts

Some Thoughts Regarding Dr. Wilson Wheatcroft's Overview of Orang Pendek Evidence.
(Originally posted to the Bigfoot Encounters website February 2008; currently on the Sasquatch Canada website.)

Dr. Wheatcroft warns that his article, "Orang Pendek: The Little Bipedal Hominid of Sumatra—An Anthropological Overview," (quote:) "may appear controversial to any reader." To me it appears quite logical, just and informative. With the exception of two points, discussed later, I find this paper very useful and welcome; first, because it updates our information on Orang Pendek, and, second, because it takes to task "the Establishment scientists," who are "elitist academic, self-proclaimed experts," and "the status quo group-think," more strongly than they have ever been taken before for ignoring the subject.

An aside in this connection; what makes a scientist? Besides having professional knowledge and training, a scientist is no scientist without the intellectual courage and curiosity to seek the truth. I do understand that demand on a scientist's courage may be excessive if his or her finding and stating the truth are fraught with the risk of losing employment. But there are thousands of zoologists and anthropologists in the world who are not facing such a risk—I mean retired scientists. Have all of them lost courage and curiosity because of age? Improbable. So why don't some of them raise their voices for the reality of bigfoot and Orang Pendek the way Dr. Wheatcroft has voiced? I think it is because they are misinformed by the tabloids and the academic self-proclaimed experts. How to open their hearts and minds? Just let them read Dr. Wheatcroft's paper. One of my intelligent bigfoot contactees (who prefers to maintain anonymity) appraised his article with the words: "a ray of light, a ray of hope." I wonder if she sees the light and hope the way I see them.

I fully agree with Dr. Wheatcroft that bigfoot and Orang Pendek are hominids (in the classification still widely used), not apes, because both are bipedal. And reading this article I realized once again with regret that we have far less evidence and knowledge about the little hominids of Sumatra and their counterparts in other habitats than about the giant hominids of North America. To borrow the term used by Tony Healy and Paul Cropper in their excellent book *The Yowie*, there are two clear-cut types of non-sapiens hominids on Earth today: littlefoot and bigfoot. Inevitably, there are also specimens and populations in between.

The author says he is uncomfortable with Ivan Sanderson's name "proto-pygmies." I am also uncomfortable if it means, or is taken to mean, that sapiens pygmies originate from non-sapiens littlefoots. This shouldn't be so. Size is the most natural and likely character for differentiation and speciation of life forms, and as a result giants and pygmies appear independently in different taxons and at different stages of evolution. Note that there are pygmy elephants and pygmy hippos. Tigers are giant cats and common cats are "pygmy tigers." Note how quickly man created pygmy (and giant) dogs and pygmy horses. If there are pygmy chimps and pygmy humans, why shouldn't there be pygmy non-sapiens hominids? As a matter of fact, reports, rumors and folk tales of them are as widespread as those of their bigger relatives. Besides Southeast Asia, they are reported in Australia, Africa, Europe, and the Americas. Why then is so little known of them? This is most probably because, due to their size, they are even more elusive and cryptic than giant hominids.

To quote the paper, "There has to date (January 2008), been no reliable photograph taken of the body, or face of the Orang Pendek, in spite of considerable efforts to get pictures..." The veteran investigator Deborah Martyr once sighted a littlefoot at a distance of 30 meters. Says she: "I had a camera in my hand at the time, but I dropped it; I was so shocked!" This is a good example of the human factor in our research.

As for the possible link between Orang Pendek and *Homo floresiensis*, I readily accept this hypothesis.

I concur with the author that much more information can be received from the eyewitnesses, provided they are properly interviewed. The leading almasty investigator, Marie-Jeanne Koffmann, devised a special questionnaire for this purpose, consisting of some 20 concrete questions. We also have special instructions for taking photographs of homin footprints (e.g., it's a must to include a scale in the photo).

Being sure that Dr. Wheatcroft is not a young man, I am impressed with the news that he "hiked in the jungle for over 9 days, with native people from two rain forest areas, in January 2007."

Now some remarks concerning the author's following statement: "This article, for the first time, has gathered all known historical as well as contemporary evidence..." Actually, it lacks the most important piece of historical evidence. It has to be realized that we are not discovering, but re-discovering relict hominids. They were first described by early naturalists; mentioned by philosophers in the Middle Ages; and finally classified as *Homo troglodytes* (*nocturnus, sylvestris*) by Linnaeus in 1758 in his *Systema Naturae*. He based this mainly on the strength of the ancient sources and on the 17th century works and reports of Dutch naturalists traveling and exploring in Southeast Asia, called at the time the East Indies. I see references to these sources in *Systema Naturae*, both in the Latin original and in a Russian translation of 1804. Also, they are in a Russian translation (done in 1777) of *Anthropomorpha*, a dissertation that Linnaeus dictated in 1760 to his student Christian Hoppius. The Latin original of the latter is not available to me.

A source cited by Linnaeus and most relevant to our discussion of Orang Pendek is the work *Historiae naturalis et medicae Indiae orientalis* by Jacob de Bondt, alias Jacobus Bontius (1592–1631), a distinguished Dutch physician who came to Batavia (now Jakarta) in Java in 1625 and lived there until his death. His book was written in Java and published in Amsterdam in 1658. He claims in it that he saw, apparently in captivity, anthropomorphic hairy bipeds, termed by him *Homo silvestris*, (i.e., forestman, man of the woods), one of which, a female, he drew a picture of and described it in detail.

Unfortunately, the Bontius book is not available to me, and here I give the information about the specimen, described by him, as mentioned by Linnaeus and academician Alexander Sevastianov in his work devoted to Linnaeus and published in 1804. The said female was hairy all over, except hands and face, head hair was long, fringing her face. Her features were quite humanlike. She was very shy, often sighed and wept, and running erect covered her genitals with a hand. She lived in a shelter of branches, slept on a bedding and put her head on a pillow. She was so human-like that it seemed she only lacked the capacity of speech to be called a human.

She was sent as a rarity to Europe but died on the way at the latitude of the Cape of Good Hope. (As to what happened to the corpse my sources are silent – D.B.) The Javanese say that these forest men can speak but hide this so as to avoid being forced to toil. According to Linnaeus, the troglodytes communicate by whistling, which is very hard for humans to learn, and of human language they can only learn to say "yes" and "no."

Bontius relates that many people believe these forest men are hybrids of apes and humans, but Linnaeus rejects this opinion and classifies them as an original species of man, *Homo troglodytes*.

Their difference from apes is bipedal locomotion, and the dental system devoid of diastemata (always present in apes and monkeys); the differences from *Homo sapiens* include night vision, membrana nictitans (so called 'third eyelid'), and the arms relatively longer than in humans. In size they are not taller than human nine-year-old boys. (Apparently, Linnaeus had no news of bigfoot – D.B.) They live in forests and stay by day in caves.

Additionally, Sevastianov, referring to information from a merchant who stayed for some time in Borneo, relates that the island is habitat to a forest man who greatly resembles a human. He is such a fast runner bipedally that he can hardly be overtaken. The king and great nobles often used to go hunting this animal. He has a wild appearance, his eyes are deeply set, the face is dark, and he looks fierce.

The above agrees with our present-day image of Orang Pendek and sedapa. It remarkably parallels Mr. van Heerwarden's description of a sedapa he observed during a hunt in Sumatra in October, 1923. At the same time it has to be stated that some bits of information, presented by Linnaeus regarding troglodytes, appear incorrect, which is not surprising considering the cryptic nature of these bipeds and popular myths and taboos connected with them—the difficulties quite conspicuous even today, in the age of information.

So I skip these dubious bits and return to the Bontius account, which looks consistent and valid. The question then is, why is his testimony being ignored by modern science or presented as a bad mistake of a naturalist? First of all, because his account has a marvelous catch; writing in Latin, Bontius also applied the Malay language words to the anthropomorphic animals he described. These words were Orang utan! How come? How on Earth could he have used the name of an ape for the obvious hominid described in his text and shown in his drawing—seen here? But who said that Orangutan is the name of an ape in the Malay language? According to my information, it is NOT. What it means is just "man of the woods," not "ape of the woods." That's how the usual popular names of relict hominids in different parts of the world are translated into English. According to my information, Bontius was the first to introduce the Malay name Orangutan to European languages and the scientific community. What's more, the ape orangutan is no longer found in Java, and it's an interesting question when this ape disappeared there; before or after Bontius?

In the 17th century, the Europeans were only beginning to learn about the great apes, and the first to be described were chimpanzees, though called then by different names. Probably following Bontius, scientists used at the time the name orangutan for the chimpanzees brought to Europe. The famous English anatomist Edward Tyson published in 1699 a monograph devoted to his study of a chimpanzee and titled it, *Ourang-outang, sive Homo sylvestris* (or The Anatomy of a Pygmie Compared with That of a Monkey, an Ape, and a Man). The ape we now call orangutan was captured in Bornea and became known to Europeans more than a century after Bontius, about 1780, and was scientifically described in Europe in 1798.

Now let us look at the zoological names of orangutan and chimpanzee. Orangutan—*Simia satyrus* (formerly), today *Pongo pygmaeus*. Pongo means ape (the word comes from Africa). So the Latin scientific name for orangutan means "pygmy ape." Quite some pygmy, almost on par with the gorilla! Chimpanzee is *Pan troglodytes,* i.e., "Pan cave-dweller." What a mess of misnomers, isn't it? And what strange borrowings of ancient popular names of relict homins. In search of an explanation for this quirk of nomenclature we have to look deeper into the history of science.

Most educated people are aware of two fundamental scientific revolutions in the history of modern civilization: Copernican and Darwinian. But few know that the latter was preceded by a revolutionary deed of Linnaeus. To the dismay and anger of the "establishment," this deed was tantamount to three intellectual "outrages." First, he instituted a zoological taxon, which included apes and monkeys, and called them by the name used by churchmen for their seniors—primates. Second, he placed man side by side with apes and monkeys in that taxon—the Order of Primates. Third, he invented a second species of man, *Homo troglodytes,* even though it was known to everybody from the Bible that God created a single man, Adam, and all people descend from him. *Homo troglodytes* was at the time a much greater pain in the back of the "establishment" than bigfoot or Orang Pendek are at present.

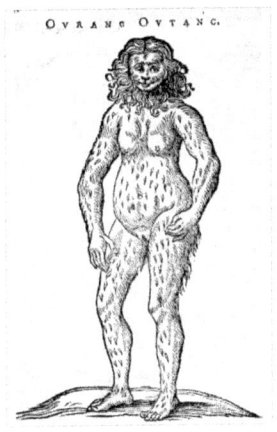

Drawing by Bontius.

In the 18th century the fame and authority of Linnaeus were so great that his most unpalatable innovations in natural history were tolerated for some time. But a backlash was inevitable. It was led by Johann Blumenbach, who in his *Manual of Natural History* (1775) established the Order of Bimanus for man and the Order of Quadrumanus for apes and monkeys. As for *Homo troglodytes,* Blumenbach discarded the species altogether as "an unintelligible mixture of pathological cases and the orangutan." He moved the term *"troglodytes"* to Simia and established *"Simia troglodytes* or Chimpansi," which implied that chimps were cave-dwellers. According to S. J. Gould, "Historical changes in classification are the fossilized indicators of conceptual revolutions." Blumenbach's monumental change in the Linnaean classification was then a conceptual counterrevolution. It lasted nearly a hundred years, until resisted and reversed by Darwin's "bulldog," Thomas Huxley (1825–1895), who with *Man's Place in Nature* (1863) restored the single Order of Primates, as well as the term itself. But *Homo troglodytes* stayed in limbo for another hundred years, until resurrected and vindicated by Boris Porshnev (1905–1972), who proclaimed yet another conceptual revolution. (Bayanov, "Some Thoughts on the Origin of Speech," 2002).

Thus, the reason for the above-mentioned terminological quirk in the nomenclature of primates was not linguistic but ideological, and its sway is felt even today. To conclude this theme and move ahead in the search for truth we have to locate the Bontius book, copy the original drawing (the copy I offer above is borrowed from Bernard Heuvelmans' and Boris Porshnev's, *L'Homme de*

Neanderthal est Toujours Vivant, 1974), as well as the relevant pages of the text and have them translated from Latin. I wish my foreign colleagues would be willing to tackle this task.

Another point in Dr. Wheatcroft's fine paper I wish to discuss is in his following words:

> ...it is erroneously assumed that "only humans have language." This is totally contradicted by studies of whales and dolphin species' exotic, high frequency languages (including communicative songs) and most high frequency, underwater sounds which are beyond the normal range of human hearing. Dolphins and whales, also, are profoundly intelligent aquatic species! Their intelligence may exceed humans, if comparative, non-verbal-based testing were to be done, objectively."

Though being no specialist on the cetaceans, I readily agree that dolphins and whales are "profoundly intelligent" animals. But if the word "intelligence" means the same to the author and me (i.e., mental capacity), I cannot agree that these animals may be more intelligent than humans. The existence of science is caused by and based on human intelligence. Humans have the science of zoology—dolphins and whales have no science of anthropology. They have no science at all. One telling result of this is that humans are exterminating whales and dolphins—whales and dolphins are not exterminating humans, and even cannot effectively defend themselves from the latter. Due to intelligence, humans are unsurpassed exterminators. As is known, with their horrific weapons they can wipe out all life on Earth. Should this happen, would the extraterrestrials say that dolphins and whales were more intelligent? I doubt this. They would probably say: "As a species, those apelike naked bipeds were not intelligent enough to be designated *Homo sapiens*."

It follows that the praiseworthy name was given our species by Linnaeus not only in contrast to *Homo troglodytes*—the caveman, but also in advance, to be proved and justified in the future. Julian Huxley said that man "is not yet fully human." Whatever the limits of our intelligence, its existence is due to the uniqueness of human language. "In the beginning (of humanness) was the Word." Words are conceptual symbols that are absent in the communications of animals, no matter how subtle and complex they may be. It is thanks to human language and its verbal symbols that Dr. Wheatcroft is able to oppose and expose the "establishment"—"Animals can 'argue' with paws and claws, but not with symbols." In consequence, the rule of animal life is status quo; the rule of human life is change and development. Animals are evolving, slowly and unconsciously. Humans, with language and intelligence (or lack thereof!) are making history, consciously and in a hurry. Language is the means not only of human existence, but also of human reproduction. Therefore there is a fundamental difference, a difference of kind, between human language and communicative systems of animals. One staunch critic of Darwin said that language is the "Rubicon of mind." That is why the question of linguistic ability of relict hominids is so crucial.

In conclusion, I want to express great appreciation of Dr. Wheatcroft's intellectual courage which is so stimulating for our research.

Postscriptum:

In June 2008, I was lucky to find a copy of the book, *Iacobi Bontii Historiae Naturalis et Medicae Indiae Orientalis*, 1658 (Book Museum of the Russian State Library, Moscow) and copied the Latin text in which Bontius describes a female specimen of a hairy hominid that he witnessed and observed in Java. The corresponding drawing by Bontius is also there, and it is the same that is reproduced from the Heuvelmans' book (previously shown).

The Latin text translation is below. It lacks some details mentioned by Alexander Sevastianov in his 1804 publication, which can mean that he had additional sources of information on this case. Nonetheless, the definite conclusion from the Bontius account and his drawing is that what he saw was not an ape, but a hominid. It follows that modern science is wrong regarding Bontius and his discovery on this account. The name "orangutan" has been wrongly applied to the big red ape, whose name in the Malay language is "mias." This means that the contribution to science by Jacobus Bontius in this regard must be recognized and the truth of the matter re-established.

Latin text translation:

Orang Utan or *Homo silvestris*.

Pliny, genius of Nature, said the following of Satyrs in Book 7, Chapter 2. There are also Satyrs in the eastern mountainous regions of India. This is a very swift animal, of human appearance, running both erect and quadrupedally. Because of its speed only old or sick can be captured. Deserving admiration, I've seen specimens of both sexes, walking erect, first a female Satyr (whose image I show here) [see page 133], very shy, hiding from unfamiliar people, weeping, covering her face with her hands, and showing other human actions which made it seem she did not lack anything human except speaking. According to the Javanese, both males and females can speak but do not want to show this so as not to be made to work. This is ridiculous. The name given them is Orang Utan which means man of forest (woodman) and it is believed they are born by Indian women whose passion makes them copulate with apes and monkeys. This is a tale not to be believed even by children either.

The text and image drawn by Bontius (page 133) indicate absolutely clearly that he claims to have seen a hairy hominid, not an anthropoid ape. This hominid had the local name of woodman which is common for homins all over the world. The name orang utan was given by Europeans to a big Asian ape by mistake. I wish the truth of the matter to become at last known to science.

The actual Latin text is as follows for those who might wish to do their own translation:

Plinius, ille Naturae Genius, lib. 7, cap. 2, de Satyris dixit: Sunt & Satyri, subsolanis in Indiis locis & montibus pernicssum animal; tum quadrupess, tum & recte currentes humana specie & effigie, propter velocitatem non nisi sense aut aegri capiuntur. Ast quod majorem meretur admirationem, vidi ego aliquot utriusque sexus erecte incedentes, imprimis eam (cuius effigiem hic exhibeo) Satyram foemellam tanta, verecundia ab ignotis sibi hominibus occulentem, tum quoque faciem minibus (liceat ita dicere) tegentem, ubertimque lacrymantem, gemitus cientem, & caeteros humanos actus exprimentem, ut nihil ei humani deesse dicers praeter loquelam. Loqui vero eos easque posse, Iavani aiunt, sed non velle, ne ad labores cogerentur: ridicule me Hercules. Nomen ei indunt Ourang Outang, quod hominem silvae significant, eosque nasci affirmant e libidine mulierum Indarum, quae se Simiis & Cercopithecis detestanda libidine miscent. Nec pueri credunt, nisi qui nondum ore lavantur.

Possible Connection with Homo floresiensis

Hominologists call the enigmatic hairy bipeds of their study, such as bigfoot, almasty, Orang Pendek, etc., relict hominids or hominoids. Relict (or relic) in biology means an organism surviving from an earlier time when it was abundant and widespread but now is rare or occurs in a small area.

Actually, the great apes—gorillas and chimpanzees of Africa, and orangutans of Asia—have become relict species at present. Most of us are aware of the famous relict fish, the coelacanth, whose fossils were discovered first, and it was believed to have gone extinct at the time of dinosaurs. In 1939, it was caught alive in the Indian Ocean and became known in the flesh to scientists. The latter also learned that local fishermen used to catch this fish from time to time and sell it on the market. The coelacanth was called a "living fossil." So in discovering this organism paleontologists were ahead of zoologists. It seems that by discovering *Homo floresiensis* in 2003 on the island of Flores in Indonesia, paleoanthropologists have come ahead of hominologists in proving the existence of "littlefoots" in this part of the world in a geologically recent past. As it follows from the Wikipedia information cited below, hominologists, in their turn, have a chance to prove the existence of littlefoots in the flesh.

Homo floresiensis ("Flores Man"; nicknamed "hobbit" and "Flo") is an extinct species in the genus *Homo*.

The remains of an individual that would have stood about 3 feet (0.91 m) in height were discovered in 2003 on the island of Flores in Indonesia. Partial skeletons of nine individuals have been recovered, including one complete cranium (skull). These remains have been the subject of intense research to determine whether they represent a species distinct from modern humans. This hominin is remarkable for its small body and brain and for its survival until relatively recent times (possibly as recently as 12,000 years ago). Recovered alongside the skeletal remains were stone tools from archaeological horizons ranging from 94,000 to 13,000 years ago. Some scholars suggest that the historical *H. floresiensis* may be connected by folk memory to Ebu Gogo myths prevalent on the isle of Flores. (...)

Additional features used to argue that the finds come from a population of previously unidentified hominids include the absence of a chin, the relatively low twist of the arm bones, and the thickness of the leg bones. The presence of each of these features has been confirmed by independent investigators but their significance has been disputed. (...)

Recent survival

The species is thought to have survived on Flores at least until 12,000 years before present, making it the longest lasting non-modern human, surviving long past the Neanderthals (*H. neanderthalensis*), which became extinct about 24,000 years ago. (...)

Local geology suggests that a volcanic eruption on Flores approximately 12,000 years ago was responsible for the demise of *H. floresiensis,* along with other local fauna, including the elephant Stegodon. Gregory Forth hypothesized that *H. floresiensis* may have survived longer in other parts of Flores to become the source of the Ebu Gogo stories told among the Nage people of Flores. The Ebu Gogo are said to have been small, hairy, language-poor cave dwellers on the scale of this species. Believed to be present at the time of the arrival of the first Portuguese ships during the 16th century, these creatures are claimed to have existed as recently as the late 19th century.

Gerd van den Bergh, a paleontologist working with the fossils, reported hearing of the Ebu Gogo a decade before the fossil discovery. On the island of Sumatra, there are reports of a 1-1.5m (3 ft 3 in–4 ft 11 in) tall humanoid, the Orang Pendek, which might be related to *H. floresiensis*. Henry Gee, senior editor at *Nature* magazine, speculates that species like *H. floresiensis* might still exist in the unexplored tropical forest of Indonesia.

CHAPTER 5

The Harm of Assumptions Turned into Convictions

(Originally posted to the Bigfoot Encounters website April 2010; currently on the Sasquatch Canada website.)

Ever new discoveries in paleoanthropology are exciting and welcome. The role of this science in humanizing and educating mankind is of special importance. Yet paleoanthropologists' total focusing on fossils and ignorance of evidence obtained by hominology is causing a great disservice to science.

Note the current items in the press:

> —If we accept that the Indonesian hobbits are yet another distinct species—and the relevant community seems to be leaning that way—then it appears that there were at least four distinct hominin species cohabiting the globe in the very recent past.

> —"We weren't alone," said Todd Disotell of New York University, who was familiar with the new work. "When we became modern, we didn't instantly replace everybody. There were other guys running around who survived quite well until very, very recently."

> —"We think it's normal to be alone in the world as we are today," Dr. Tattersall said, and to see human evolution as a long trend leading to *Homo sapiens*. In fact, the tree has kept generating new branches that get cut off, presumably by the sole survivor. "The fossil record is very eloquent about this, and it's telling us we are an insuperable competitor," Dr. Tattersall said. Modern humans' edge over other species probably emerged from their ability to process information: "We can invent alternatives in our heads instead of accepting nature as it is," Dr. Tattersall said.

No, Dr. Tattersall, it's NOT normal for *Homo sapiens* to be alone in the world, if you accept evolution. In evolutionary terms, there is no less, perhaps more, reason for non-sapiens hominins (hominids) to be with us today than for chimpanzees and gorillas. Being so sure of *Homo sapiens* survival ability, are you being logical in denying this ability to our close hominid relatives? When we became "modern," we not only didn't instantly replace "other guys running around" but worshiped them as gods over tens of millennia because of their edge over us in the wilderness. The ever repeated assumptions, turned into convictions, of the other hominids total extinction are based on IGNORANCE, not facts and knowledge.

Paleoanthropologists ignore Darwin's views on the question of species extinction:

> No fixed law seems to determine the length of time during which any single species or any single genus endures," and "the utter extinction of a whole group of species has sometimes been a slow process, from the survival of a few descendants, lingering in protected and isolated situations." (*The Origin of Species by Natural Selection*, 1929, pp. 280 and 299).

They ignore the views of a prominent paleontologist on the same question:

> It is always necessary to remember the incompleteness of the geological record. The first appearance of a given species in the geological record and its disappearance from the latter can in no way be taken for the dates of its origin and final extinction. The real life span of a species (or a group of species) is usually much longer than the period determined from the geological record. Consequently, the dating of the extinction of a form or a group is not as simple a matter as may appear from the frequent citing in the paleontological literature of extinction dates for various organisms." (*History of Evolutionary Paleontology from Darwin to Our Days* [in Russian], L.S. Davitashvili, 1948, p. 486).

They ignore the lesson of the coelacanth survival, the fish formerly known only from the fossil record and thought to have been extinct for sixty to seventy million years.

They ignore the origin of the terms *Homo sapiens* and *Homo troglodytes,* that is the Linnaean contribution to anthropology and his founding of primatology, which was banned after his death. In 2003 I wrote:

> One of the great scientific results of the 20th century was the discovery of relict hominids (homins, for short), popularly known as abominable snowman, yeti, yeren, almas, bigfoot, sasquatch, etc. Actually, it was a re-discovery by hominologists of what had been known to western naturalists from antiquity to the middle of the 18th century, when wild bipedal primates were classified by Carl Linnaeus as *Homo troglodytes* (i.e., caveman) or *Homo sylvestris* (i.e., woodman). As for eastern scholars and rural populations in many parts of the world, they have always been aware of wild hairy bipeds, known under diverse popular names.

Paleoanthropologists, along with the rest of the scientific community, ignore the abundant evidence, including a documentary film, put together over half a century ago by hominologists, whose work is "blessed" with academic cover-ups. Specialists, ignorant of hominology, have usurped the role of judges on these matters in scientific journals and mass media, which is detrimental for truth and progress in science. But truth will get out, and the shock and derision these judges will get will be well-deserved. With due respect for their discoveries in the ground, paleoanthropologists will suffer for their divorce from life on the ground.

CHAPTER 6

Thoughts on the Revolution in Anthropology

(Originally published on the Bigfoot Encounters website February 2011; currently on the Sasquatch Canada website.)

I. What is Holding the Revolution Back?

I was moved to write this paper by Dr. John Bindernagel's book *The Discovery of the Sasquatch*, 2010. The author is a Canadian biologist with over forty years of experience in wildlife research and conservation and an active investigator of the sasquatch/bigfoot phenomenon. Dr. Bindernagel writes that discovery in science is a process and must take time:

> But even acknowledging discovery as a process, the discovery of the sasquatch may appear to have been abnormally prolonged. (p. 145)

> One reason that discoveries become prolonged is that they may have been neglected or ignored, in which case *re*discovery may be necessary before they are finally acknowledged. (p. 127)

> For the few scientists with "relevant research agendas," the patterns of great ape anatomy and behavior and the capability of the great ape hypothesis to explain the evidence suggest the possibility that the sasquatch has already been discovered, though it has not yet been officially cataloged. (p. 128)

Final words from the Epilogue:

> It is hoped that the attempted reconciliation of the diverse facets of this unique discovery process as presented here, will contribute not only towards the completion of the discovery process, but, in addition, to a greater understanding of its prolonged nature. (p. 236)

To my mind, for a greater understanding of the whole phenomenon, discovery and rediscovery, on one side, and their acknowledgement, on the other, must be clearly set apart. Yes, discoveries may be neglected and followed by rediscoveries.

The motion of the Earth was stated by Aristarchus of Samos, 3rd century B.C., and "rediscovered" by Copernicus in the 16th century A.C. The acknowledgement process then took over a hundred years and became known as the Copernican revolution. The origin of the science of meteoritics is closer to our situation. Stones falling from heaven have always been known to people around the world, but not always to scientists. So it cannot be said that meteorites were discovered in the 18th century, when their nature as such was acknowledged by science in a process of several decades. "Wild men", i.e., wild hairy bipedal primates, have also been known to people around the world throughout history. So who can literally claim their discovery? It can be said that the process of their rediscovery by Western scientists began in the 1950s with the yeti Himalayan expeditions and ended with the validation by hominologists of the Patterson/Gimlin documentary film. Thus, as a hominologist, I can firmly state today that wild bipedal higher primates, i.e., hominids different from *Homo sapiens,* still inhabit all continents, with the exception of Antarctica. That this fact is not acknowledged by non-hominologists is a totally different matter, and it is this that calls for clarification and explanation.

One of the most important works on the process and progress of science is Thomas Kuhn's book, *The Structure of Scientific Revolutions,* 1962. Key terms and notions introduced by Kuhn are **normal science, revolutionary science** and the **paradigm shift**. Bindernagel cites Kuhn on several pages, but only in regard to normal science; not a word is to be found in his book about revolutionary science and the paradigm shift. For example, he offers these quotes from Thomas Kuhn's book:

> Normal science, the activity in which most scientists inevitably spend almost all their time, is predicated on the assumption that the scientific community knows what the world is like. (...)
>
> Normal science, for example, often suppresses fundamental

novelties because they are necessarily subversive of its basic commitments.

In contrast, my teacher of hominology, Professor Boris Porshnev (1905-1972), referred to Thomas Kuhn's book and spelled out scientific revolution in the very first sentence of the article "Is a Scientific Revolution in Primatology Possible Today?" The article was published in 1966 in the Russian journal, *Questions of Philosophy*, issued by the Institute of Philosophy under the auspices of the Soviet Academy of Sciences. Porshnev said in his article that the information about so-called relict hominoids ("abominable snowmen") that surfaced at the time, and was analyzed by him in his volume, *The Present State of the Question of Relict Hominoids*, 1963, could not be explained and understood without a paradigm shift—therefore a scientific revolution was the order of the day.

Porshnev knew well what he was doing—a Scientific Revolution—and therefore he was a revolutionary in science. Regrettably, I am not aware of any other professor or PhD scientist in the world who could be called a conscious and deliberate revolutionary in our field of research. The only other scientist who foresaw, albeit reluctantly, a tectonic transformation in store for primatology and anthropology was primatologist and paleoanthropologist Dr. John Napier (1917–1987). In his book, *Bigfoot: The Yeti and Sasquatch In Myth and Reality*, 1973, he wrote that if bigfoot was real,

> —then as scientists we have a lot to explain. Among other things we shall have to re-write the story of human evolution. We shall have to accept that *Homo sapiens* is not the one and only living product of the hominid line, and we shall have to admit that there are still major mysteries to be solved in a world we thought we knew so well. (p.204)

Note the words that the reality of bigfoot would mean "that *Homo sapiens* is not the one and only living product of the hominid line," which invalidates John Bindernagel's great ape hypothesis to explain the bigfoot/sasquatch phenomenon. The discovery of an uncataloged great ape would not make a revolution in science and would not make anthropologists re-write the story of human evolution. The discovery of a living hominid different from *Homo sapiens* would certainly have such an effect, because it's a long held dogma of anthropology that *Homo sapiens* is "the one and only living product of the hominid line." As I argued at length in my paper, "Is a Manimal more Man than Animal?," 2005, the great ape hypothesis is inconsistent and misleading for our research. To date no counter-arguments have been offered. Unfortunately, the names of such books as, *Sasquatch: The Apes Among Us*, 1978; *North America's Great Ape: the Sasquatch*, 1998; *Bigfoot! The True Story of Apes in America*, 2003, authored by the leading hominologists of North America, present bigfoot/sasquatch as real, not hypothetical, apes.

As for John Napier, being a typical worker of normal science, he tried to blacklist all evidence of our revolutionary science, including the Patterson/Gimlin documentary film. So it's the depth of transformations in anthropology, the perspective of an actual revolution in science, which is the main cause of the "abnormally prolonged" delay of official acknowledgement of the existence of sasquatch and other non-*Homo sapiens* hominids. The scope and tactics of the counter-revolutionary resistance by academics was not foreseen even by Boris Porshnev. His provocative paper published in a philosophy journal was printed under the rubric "For discussion," and he sincerely looked forward to reading the opinions of his peers. Their reaction was unprecedented in the history of the journal; complete silence and neglect of the professor's challenge. Ever since, this tactic and reaction of the academic circles have become predominant in regard to hominological investigations. This is not to say that direct personal attacks have been lacking. Four zoologists, in an article in their academic journal, accused Porshnev of spreading pseudoscience, hinted that his mind was abnormal, and asked whether people "circulating such yarns have the right to bear the honorary title of a Soviet scientific worker." (*Vestnik Zoologii*, 1969, No. 4, pp. 69-80). Porshnev's ideas on the subject then became, and still remain, taboo in the Russian academic circles.

As I said, "wild men" were rediscovered by Western scientists in the middle of the last century, and the credit for rediscovery, on a theoretical and scientific level, definitely goes to Professor Porshnev. It is in his "semi-secret" volume of 416 pages, *The Present State of the Question of Relict Hominoids*, that he named and honored his predecessor in the research, Linnaeus (1707-1778). I call the book "semi-secret" because there existed only 180 copies printed for the Soviet Academy of Sciences high officials, so its contents were unknown to most scientists in the world. (**Note:** 2,500 copies of the work were published in 2012.)

Thanks to my acquaintance with the Professor in 1964, I read the book and became riveted to the subject. As to Linnaeus, he was as great a natural science celebrity in the 18th century as Charles Darwin was in the 19th century. Linnaeus established the binominal system of designation of plants and animals, and it was said at the time, "God created things, Linnaeus put them in order." All educated people are aware of two fundamental scientific revolutions in the history of modern civilization: the Copernican and the Darwinian. But few know that the latter was preceded by a revolutionary deed of Linnaeus. To the dismay and anger of the "establishment," this deed was tantamount to three intellectual "outrages." First, he instituted a zoological taxon, which included apes and monkeys, and called them by the name used by churchmen for their seniors—Primates. Second, he placed man side by side with apes and monkeys in that taxon—the Order of Primates. Third, he "invented" a second species of man, *Homo troglodytes,* when it is known from a creationist standpoint that God created a single man, Adam, and all people descend from him. *Homo troglodytes* was at the time a much greater pain in the backside of the "establishment" than bigfoot is at present.

Also known only by few is the fact that it was Linnaeus who introduced in science the central term of anthropology—*Homo sapiens*—and did so a century before the discovery and study of fossil hominids. Nobody wonders today why man was given such an incongruous scientific name. Well, as mentioned above, the Linnaean nomenclature, published in the 10th edition of his *Systema Naturae* (1758), included not one but TWO living species of man: *Homo sapiens* (man the wise) and *Homo troglodytes* (the caveman). Importantly, the latter term was not coined by Linnaeus—he borrowed it from ancient naturalists, and he described *Homo troglodytes* as *nocturnes* (nocturnal), and *sylvestris* (of the forest), two characteristics ringing a bell for all hominologists. So there is no doubt that our kind owes its undeserved name of "man the wise" in contrast to the "caveman" in the Linnaean classification.

Linnaeus based his description of troglodytes on the writings of ancient naturalists and the accounts of travelers of his epoch. Inevitably, his information was patchy and contradictory, which made him write in the dissertation, *Anthropomorpha* (1760), that he dictated (which was usual at the time) to his student Christian Hoppius:

> Is it not amazing that man, endowed by nature with curiosity, has left the Troglodytes in the dark and did not want to investigate the creatures that resemble him to such a high degree? (...) As for me, I remain in doubt what specific characteristic distinguishes the Troglodyte from man (*Homo sapiens* – D.B.) within the scope of natural history.

The questions posed by the great man of science are as relevant today as they were in his time (as I have detailed on page 135).

Every hominologist worth his salt must know the history of primatology, and the reason why it was not all sweet and easy. The problems of hominology would not seem so inscrutable were this so. As a matter of fact, most people don't even know they are also primates. The best indication of that, and that our animal ancestors were arboreal, are dermal ridges on our palms and soles—the hallmark of primates; dermal ridges are also detected in some clear sasquatch footprints.

When I am asked why living *Homo troglodytes L.* was not known to science after Linnaeus and until rediscovered by Porshnev, I reply that: it was not known to science because there was no science to know it—I mean a <u>natural, biological</u> science. This is convincingly demonstrated by the case of the Russian zoologist, Professor Vitaly Khakhlov, who in 1914, as a college student, collected information on the wild man in Central Asia, named it *Primihomo asiaticus,* and reported his findings to the Russian Academy of Sciences. In the 1960s, his report was dug up by Professor Porshnev in the Academy's archive from the file labeled "Notes of no scientific significance."

And why did the living wild man stay in limbo in a century-long period between Thomas Huxley and Boris Porshnev? Mainly there were two reasons as follows.

First, at the time, Neanderthal and later other fossil hominids had been discovered, and the science of paleoanthropology, aimed at finding hominids <u>in the ground, not on the ground</u>, began coming into its own. Its origin and history were not straight and easy (there were big problems and delays with the acknowledgement of early Neanderthal, *Homo erectus*, and *Australopithecus* fossils), which tended to make scholars of this discipline rather self-centered and "looking down more than around." Paleoanthropology has since played, and is playing, a great and unique role in promoting man's self-knowledge, and it is a very respectable science today. The paleoanthropologist's inward attitude to our subject I tried to encapsulate in one chapter heading of my work, "A Hominid Fossil in the Hand is Worth Two Homins in the Bush." Actually, paleoanthropology is the elder sister of hominology, but the latter's treatment by the former is much sterner than was the treatment of Cinderella by her sisters.

The second reason was that *Homo troglodytes L.*, alias wild man, is one of the main, if not the main, hero of folklore, mythology and demonology the world over. Naturally, it was and is well-known to humanitarian disciplines of knowledge, such as folkloristics and the study of mythology and demonology, but known not as reality but as myth. And this was the greatest impediment for the birth of hominology, and remains so for its acknowledgement at present.

The opposition charged Porshnev with preaching pseudo-science, asserting that he took for reality a figment of the imagination and mythology. Porshnev's main opponent, Professor Nikolai Vereshchagin, an outstanding zoologist and paleontologist, was quoted in the weekly *Moscow News,* No. 42, (1979), as follows:

> My opinion is that while legends about trolls, demons and witches have lost their credibility with modern Europeans, travelers and mountain climbers have probably fallen hook, line and sinker for similar legends and myths current among the peoples of the Himalayas and the Pamirs, giving enthusiasts the fuel they desire.

I wrote back in 1976:

> If in the course of history, people had encounters with

"troglodytes," then these most impressive beings could not have escaped the attention of the creators of myths and legends. (...) Is the abundant folklore, say, about the wolf or the bear not a consequence of the existence of these animals and man's knowledge of them? Therefore, we say that if relic hominoids were not reflected in folklore and mythology, then their reality could be called into question. Fortunately, this channel of information is so wide and deep that much work can be done in this sphere: it is necessary to re-examine and re-think a good many anthropomorphic images playing important roles in folklore and demonology.

In 1991, thanks to Gorbachev's *perestroika,* which gave people the freedom of speech and press, I was able to publish at last my book in Russian, *Wood Goblin Dubbed Monkey: A Comparative Study in Demonology.* I re-examined and re-interpreted in it a great many anthropomorphic images from the ethnic folklore and demonology of numerous peoples of the Soviet Union. It laid bare the identical biological basis in all of them, revealing the physical appearance, behavior and interactions with humans of various "demons." The book was published right on the eve of the Soviet Union's collapse, with no soul giving thought at the time to the nature and fate of wood goblins and the like. Subsequent life in Russia has not been easy either, and as a result the work's contents and message still remain unknown to the scientific community—with one notable exception. As a gesture of defiance, I sent a copy of the book to Professor Vereshchagin, with the inscription: "Greetings from the wood goblins of the 20th century!" I never expected a reply. To my amazement, it did come, with much praise for me and the book. As a result, the snowman's foremost foe changed his mind and accepted the reality of "wood goblins," as it transpired from our further friendly communications. He died in 2008, at the age of 100 less a month. I pride myself on the conversion of at least one bitter critic into a friend and supporter.

Still the "mythology barrier" is as high as ever for hominology. A number of opponents in the West claim that hairy wildmen are legendary creations not only of illiterate people of past centuries, but also of modern educated citizens. American anthropologist, Dr. David Daegling, is one such critic, having written the book, *Bigfoot*

Exposed: An Anthropologist Examines America's Enduring Legend, 2004. His final conclusion reads: "There is a mystery here, to be sure, but it is not deep within the forest where the answer lies." All deliberations of the author are aimed at proving that the answer lies in people's imagination. The book has received much praise from the author's learned colleagues, as quoted on the back cover as follows: "The book is terrific!" and "David Daegling has written a wonderful book on the North American Bigfoot myth." Dr. Daegling's distortions of facts and wrong conclusions are thoroughly examined and exposed in my book, *Bigfoot Research: The Russian Vision,* 2007, 2011.

Now back to Porshnev and his predecessors. His second great forerunner, after Linnaeus, was Darwin himself. Linnaeus lived and worked in the pre-Darwinian epoch, Porshnev in the post-Darwinian. Linnaeus was, in accordance with the paradigm of his time, a creationist. Porshnev was a devoted evolutionist and Darwinist. Hominology owes its birth and origin to evolutionary theory and the Darwinian revolution. Actually, hominology is a follow-up and continuation of that revolution. Darwin's name and contribution to science are known better than those of Linnaeus, so I won't dwell on them here, except to mention one episode of Darwin's biography directly connected with our subject.

In the year 1833, during the course of his famous five-year voyage around the world on the *Beagle* (1831–1836), Charles Darwin was very surprised with the evidence of wildmen, suddenly encountered at "the end of the world," in Tierra del Fuego, the southern part of South America. He was told by a Fuegian, who served as a guide of the expedition and was called York by expedition members, that his brother had once killed a "wild man" during a hunt. Darwin writes in Chapter X of *A Naturalist's Voyage Round the World*:

> What the "bad wild men" were has always appeared to me most mysterious; (this implies that he heard more about wild men than is mentioned in his account — D.B.) from what York said, when he found the place like the form of a hare, where a single man had slept the night before, I should have thought that they were thieves who had been driven from their tribes, but other obscure speeches (what a pity they are not related — D.B.) made me doubt this; I have

sometimes imagined that the most probable explanation was that they were insane.

The last sentence indicates that Darwin gave much thought to the mystery. As I wrote in 1984:

> I am much inclined to think that the creatures described as "wild men" by the savages of Tierra del Fuego were not *Homo sapiens* but *Troglodytes recens ubiquitous*. Realizing that Darwin himself may have been close to a live object of our long and tortuous research, undertaken in the light of his great and revolutionary theory, I can't help feeling sort of elation mixed with wonder. It is intriguing to conjecture what course anthropology might have taken had Darwin happened to see the "bad wild man" whose sleeping place he was shown."

I hope the reader is now prepared enough to understand why scientific acknowledgement in our case has been "abnormally prolonged." It is said, "Lasting changes come slowly" (Halton Arp). In fact, considering the revolutionary nature of our endeavors, the delay is not abnormal, but normal. History of science teaches us that it is normal for normal science to neglect and then resist revolutionary science as long as possible. The paradigm shift is a painful process for orthodoxy. As John Darnton put it, with over-emphasis, "Science will turn to superstition and torture to defend its right to be wrong" (*Neanderthal*, p. 57). It is our behavior which is abnormal, not that of our opponents. It is abnormal because nearly all hominologists don't even regard themselves as such—in particular the few scientists with "relevant research agendas," (as Dr. Bindernagel elegantly put it). They never come up with revolutionary agendas, never say they are making a revolution in science— perhaps not even being conscious of it. They behave like workers of normal science, making their careers and bent on personal, not collective efforts and goals. No wonder each bigfoot researcher, even with academic credentials, is regarded as a maverick and ignored by the scientific community. Scientific revolutions are not gaining speed in this way. This is our own considerable contribution to the delay of scientific acknowledgement.

How do we know that Neanderthals and other fossil hominids are fact, not fiction? We know this because there is a scientific discipline called paleoanthropology, with its specialists, paleoanthropologists. Actually, all things and subjects in science are accepted as reality thanks to corresponding disciplines and specialists. It takes a specialist of meteoritics to tell a stone fallen from heaven from a stone belonging to Earth. It takes a specialist of a different kind to tell a hairy wild man from a hairy human (a case of hypertrichosis). Scientific disciplines are usually instituted by considering two criteria: specificity of study subjects, and their relative importance. Thus, paleoanthropology split from paleontology and was instituted as a separate discipline on account of hominid specificity and the subject's importance for understanding man's origin. Primatology was instituted as a separate discipline within zoology on account of primate specifics and importance of this order of animals. Hominology is singled out and instituted by the same token. Its study subjects are living hominids, which makes hominology different from paleoanthropology. As these hairy bipedal primates are different both from apes and *Homo sapiens,* their study can't be but a separate and specific field of knowledge. To know and remember this is essential for correcting our abnormal tactics and strategy on the way to acknowledgement.

All members of the scientific community know the words "abominable snowman" and "bigfoot," but they don't know a scientific discipline called hominology. If they knew it and accepted it as scientific, they wouldn't fail to take its subjects of study seriously as well. Hence, our priority is not to get the rediscovery of living hominids acknowledged (it's futile for the moment), but to build the science of them. This means, first of all, to come to agreement on basic points among members of our own research community. Paleoanthropologists are a very contentious lot, but they come out as members of a single discipline, accepting and supporting certain common rules and principles, distinguishing them from other scientists. Similarly, all members of our research community must accept at last the name of the discipline which we represent and work for. The terms *hominology* and *hominologist* have long been in use and I hope they will be legitimized at last by general agreement.

It is essential, of course, to come to terms among ourselves on the nature and taxon status of the primates we are re-discovering. As the phenomenon is global, we must take a global approach in considering this question. Such approach was practiced by the founding fathers of our research—Ivan Sanderson and Boris Porshnev, but it was abandoned subsequently and became isolated in North America, China and Australia. It has to be also realized that our discipline cannot be properly instituted in practice without being embodied and represented by an appropriate international institution such as an international society or association. What we badly need is not a bigfoot body, but a solid scientific body. No doubt, to be taken seriously by the mainstream, we must present a united front and a single scientific current (with inevitable undercurrents, of course). A call to this end sounded back in 1970s—Hominologists of all lands, unite to show humankind what is true and right!

Finally, all our members must know and remember that hominology is not a hobby or deviating pastime, but the locomotive and beacon of a revolution in science. Below are my thoughts on the latter.

II. The Revolution's Impact and Significance

In 1975, John Green, René Dahinden, George Haas, Gordon Strasenburgh and I were discussing whether it was permissible to kill a bigfoot to prove the beings' existence. George Haas and I were against the idea. He was the organizer and spokesman of the Bay Area Group (bigfoot investigators in California), and publisher of the *Bigfoot Bulletin,* a new kind of venture in North American hominology at the time. His arguments and ideas greatly impressed me. He stated that all animals on the "spaceship Earth" are entitled to the respect and consideration due to any fellow traveler. In his own words:

> As individuals, as groups, as societies, we, in effect, hold all things in trust for future generations, not only of men but of all other species as well. How we manage this self-appointed trust is the measure of our integrity. If we log off all the redwood groves for the sake of a few jobs, if we exterminate all the coyotes to save a few ranchers' sheep, if we kill off all the eagles for a few souvenir feathers, then our sense of values is warped and distorted and we have failed to live up to our trust.

If people of such mentality were not an exception but the rule on what he called the "spaceship Earth," we could be fairly confident of mankind's continued long travel through space and time. Nobody is confident of that today. In all aspects and spheres—be it the economy, politics, ecology, biodiversity—conditions on Earth would be startlingly different if the majority of humans thought and behaved like George Haas. Is it Utopian to expect most people ever to think and behave in such a wise and life-affirmative way? Utopian or not, there seems to be no way out of humanity's global crisis without applying the best means of overcoming it. The best means of changing people's mentality and behavior is enlightenment and education.

Of the global army of teachers and educators, the most amazing and successful, among those known to me, is primatologist Dr. Jane Goodall, with her global project Roots & Shoots. Addressing young people, she wrote:

> My greatest hope lies in the fact that young people, all around the world, are not only aware of the problems, but actually want to try to help solve them. And because the future of the planet lies in the hands of today's and tomorrow's youth, I decided I wanted to do my share of trying to help you to help the world. My way of helping was to start Roots & Shoots.
>
> (...) It began with one small group of high school students in Dar es Salaam in 1991. (...) It is called Roots & Shoots because roots move gradually under the ground to make a firm foundation, and shoots seem small and weak, but to reach the light they can break open brick walls. The brick walls are all the problems we have been talking about. The roots and shoots are you and your friends and young people all around the world. Hundreds and thousands of roots and shoots can solve the problems, change the world, and make it a better place to live" (*My Life with the Chimpanzees*, 1996, pp. 141,142).

Some salient features of Roots & Shoots programs are these: they are addressed to and engage young people of different cultural and religious backgrounds; they encourage them to solve problems cooperatively and become sensitive, socially responsible members of the community; the students accomplish projects they design themselves; it's a way to develop critical thinking, build better minds and connect learning with real global issues.

Now visualize real global issues, with people like George Haas and Jane Goodall, with her Roots & Shoots, on one side, and young men and women, on the other side, who become, or are turned into, terrorist "live bombs" and are committing horrendous mass murders around the world. It is said that the young men, responsible for the September 11, 2001 horror in the U.S., were university educated. Were they then genetically different from the rest of us? Of course not; they were mentally different. And what built their mentality? Clearly, it was their upbringing and education at an early age, which a university education could no longer alter, even if it could alter mentality at all.

I make these points regarding education before turning to the coming revolution in anthropology because revolutions in science are opening new vistas and new chapters in the history of enlightenment and education. This is especially true of the Copernican and the Darwinian revolutions. I don't know how evolutionists will rewrite the story of human evolution when they acknowledge the reality of relict hominids, because we don't know yet the place of these primates on the tree of evolution and the number of their species or subspecies existing on the planet today. But we can make comparisons with the previous revolutions and predict certain cognitive and educational effects of the coming one.

The effect of the Copernican revolution was tremendous, but it was somewhat diluted by the long time of its acknowledgement; it affected first of all the educated upper strata of society, and initially only in one part of the world, i.e., Western Europe. It was a radical and gigantic advance in knowledge for mankind, and consequently in the worldview and world enlightenment. But I think this effect has never been used sufficiently in education to develop critical thinking and build better minds. Curiously, the thought-provoking metaphor "spaceship Earth" could only be coined in the age of astronautics, that came as a result of the Copernican revolution. Copernicus was said to have moved the Earth and stopped the sun. If we want to see the world a safer and better place to live, all young people must know the feat of Copernicus and Galileo, as well as that of the greatest hero of science—Giordano Bruno. Without such knowledge they wouldn't know their place in the universe and wouldn't feel themselves passengers on the "spaceship Earth."

The results of the Darwinian revolution are no less telling. Darwin sheds light on man's place in nature, consequently, on man's nature itself. "Know thyself" remains the highest commandment of all. Did Darwin know all about evolution? Of course not; nobody does, but he provided the theory of evolution and discovered some of its laws. The question of man's origin and nature is even more touchy for humans than the place of Earth in the universe. If Darwin had lived and come up with his theory a couple of centuries earlier, he would have been jailed, like Galileo, or burnt alive, like Giordano Bruno. Darwin's views and theory are still being vehemently attacked, so the Darwinian revolution is still going on. Richard Dawkins wrote that "Intelligent life on a planet comes of age when it first works out the reason for its own existence," i.e., the idea of evolution. As the latter is still being ignored or denied by the majority of earthlings, intelligent life on our planet has not really come of age yet. But one potent indicator that it is moving in the right direction is this: On October 23, 1996, Pope John Paul II said in a speech to the Pontifical Academy of Sciences:

> New findings lead us toward the recognition of evolution as more than a hypothesis. In fact it is remarkable that this theory has had progressively greater influence on the spirit of researchers, following a series of discoveries in different scholarly disciplines. The convergence in the results of these independent studies—which was neither planned nor sought—constitutes in itself a significant argument in favor of the theory.

If Moslem high hierarchs likewise recognized that evolution is "more than a hypothesis," if madrasah students were taught Darwinism, I bet they would not be recruited to commit mass murders in the name of God. This shows how closely science and enlightenment are connected both with daily life and global issues.

Now to the revolution in anthropology. In comparison, the issue is no big deal. Hominids thought to be extinct turn out to be extant.

Yet a cognitive and emotional shock received from the news of their confirmed existence by the world of science is likely to be greater than in the previous revolutions. The reason is in the modern means of information that will bring the news in no time to every radio listener and TV viewer on the planet. bigfoot and yeti are of more interest to common people than abstractions of cosmology and evolution, so billions of listeners and viewers will expect explanations from the spokesmen of the scientific establishment—from the people who for decades have been treating the subject with naysayings and ridicule.

Roger Knights, a Washington State reseacher, proposed to build a wall of shame with their names on it. A good idea, but I am more interested in something else. When our academic opponents have eaten enough humble pie and finally recognized the presence of "uncataloged" hominid passengers on the "spaceship Earth," there will be a marvelous event in the history of science—natural sciences will learn a historic lesson from humanitarian sciences. In other words, biologists, paleoanthropologists and physical anthropologists will take a great lesson from cultural anthropology, from folkloristics and mythology. It is these fields of knowledge that have preserved the bulk of evidence for the existence of uncataloged hominids and the evidence stubbornly ignored and denied by the learned skeptics. And it will be a historic lesson taken by science and scientists from lay people, and from the native population all over the world who have always known the presence of hairy wildmen—which some have made no secret of.

I must say that scientists (many of them) are in need of such lessons. They make a fetish of their discipline, they worship paradigms of science as if these were dogmas of religion. Science is great and marvelous, but it has picked up and consumed only crumbs of the infinite *truth,* called the universe. So the scientist is always in need of Socratic humility and in need of always repeating, "I know that I don't know." In the case of hominology, a wide and surprising gap between popular knowledge and scientific knowledge (shall I say ignorance?) will be filled up and terminated at last. But who really knows whether or not there are other gaps of this kind and in what numbers? It's appropriate here to repeat Thomas Kuhn's observation: "Normal science, the activity in which most scientists inevitably spend almost all their time, is predicated on the assumption that the scientific community knows what the world is like." That the assumption is wrong or of limited worth follows from the words of John Haldane: "The Universe is not only queerer than we suppose, but queerer than we can suppose."

Surely, acknowledgement of hominology and the revolution in anthropology are bound to have a sanative and stimulating effect on the world scientific community. This in turn must enhance the role of science in the enlightenment of peoples, in their increased taste for democracy and choice of better leaders for government. Popular masses, scientists and governments will then learn that the redwoods, the coyotes and the eagles, just like bigfoots, yetis and all our relatives of this kind, need to be given due consideration as our fellow travelers on the "spaceship Earth." That's what George Haas called for and hoped for.

Spaceships that spin around the planet are held in orbit not only by their speed and gravity, but also, and even more so, by the skill, intelligence and team-work of their crews. Humans have to behave and manage the planet in the same intelligent and cooperative way in order to insure humanity's continued travel through time and space.

Color Photograph Presentation

The following photographs show objects of interest, some of my travels, some of the many great people I have met on my long journey, my family, the city of Moscow (in which I live), and my little country retreat.

It is hoped that the presentation will assist the reader to better understand my passion for hominology and my sincere hope that this all but ignored discipline will become a recognized branch of science.

Shown here is a fragment of an old Russian icon from the collection of the Museum of the Moscow Kremlin. Called The Virgin Bogolyubskaya, it shows scenes from the lives of Saints Zosima and Savvati. It was painted in 1545 at the Solovetsky Monastery in the north of Russia. The scene shows the hermits Zosima and Savvati being tempted by the devil. The latter is portrayed in the image of a shaggy biped better known today in Russia by the name of "snowman." The depiction of such creatures in ancient and medieval art is ubiquitous and very instructive. Deified and worshiped as lords of nature in heathen times, they were subsequently condemned and turned into demons in the major religions—Zoroastrianism, Judaism, Christianity and Islam. The demonological and religious connections of snowmen, alias relict hominoids, have for ages camouflaged their true nature and prevented science from investigating the question in earnest. The situation is just beginning to change. (Photo: D. Bayanov)

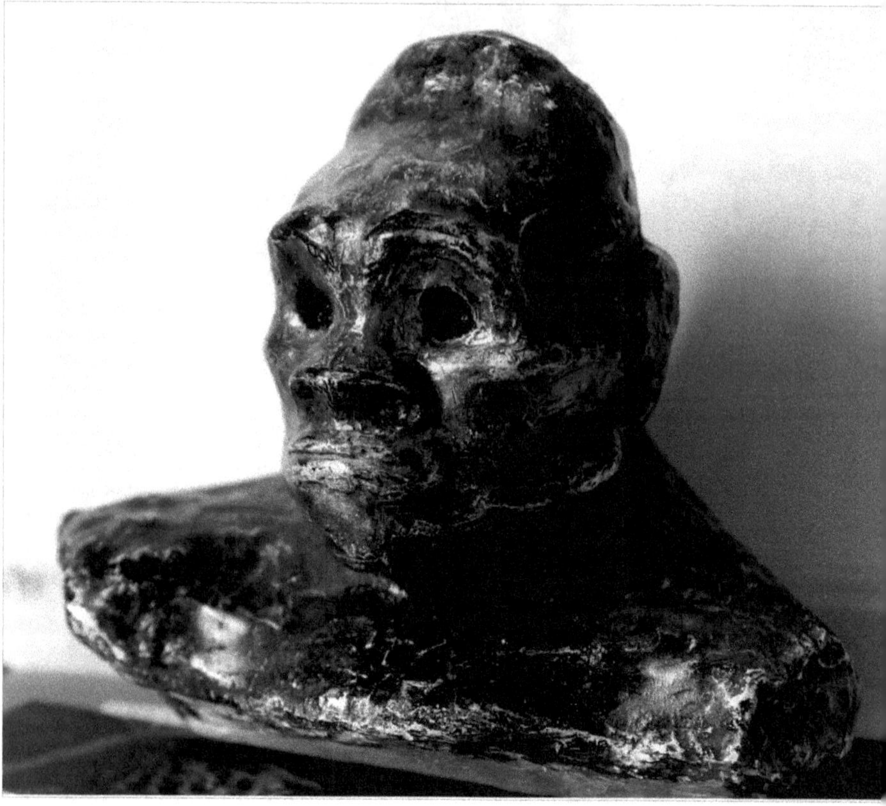

A sculpture providing two views of "Patty," the sasquatch seen in the Patterson/Gimlin documentary film, by Alexandra Bourtseva. Alexandra created the work in 1974 and gifted it to me as a birthday present that year. The protrusion on Patty's head was caused, as interpreted by Russian analysts, by the mass of head hair, not a sagittal bony crest. (Photos: D. Bayanov)

Lydia Bourtseva, a talented Russian artist, is seen here with her interpretation (rather flattering in my view) of the sasquatch seen in the Patterson/Gimlin documentary film. Lydia has created many illustrations pertaining to the sasquatch and almasty. She designed the cover for my book, *BIGFOOT: To Kill or To Film—The Problem of Proof,* and her depiction of Dr. Karapetian and the hairy man appears in Chris Murphy's book, *Know the Sasquatch* (page. 279) (Photo: I. Burtsev)

Right: Author during the 1982 expedition in Tajikistan.

Below: Perched precariously on slanting rock, the author demonstrates weathered bones that brought him and Vadim Makarov half way across the country to the Tien Shan Mountains. The find had been reported by a local hunter. We identified the skeletal remains as those of *Homo sapiens*. (Photos: D. Bayanov)

Upper left: View of the location of a hominoid encounter in 1980 by expedition member Nina Grinyova; Hissar Range, Tajikistan. (Photo: D. Bayanov)
Upper right: Photo to illustrate the availability of hominoid food in the Chukchi Peninsula. (Photo: Alexandra Bourtseva)
Left: Beautiful Lake Pairon in the Karatag Gorge of the Hissar Range in Tajikistan. It was the place of a female hominoid sighting in 1980 by two members of our hominology seminar at the Darwin Museum. (Photo: D. Bayanov)

Left: The steep rockface on the bank of the Chusovaya River in the Urals area, where witness Alexander Katayev sighted in 1974 two homins, male and female, who swam across the river and climbed the rockface "very quickly." (Photo: Author's file)

Below: A summer camp Kazakh cattle-breeders Makarov and I visited on our way to the cave with skeletal remains. The hosts offered us a good meal but were reticent regarding the wild man subject. (Photo: A. Katayev)

ght: Author and Editor, in happy unison, ossing Oregon by car en route to the llow Creek Bigfoot Symposium in ptember 2003. I was lucky to cross ths with Chris Murphy in the early 1990s. e have since worked together on a number of projects, including this book. (Photo: Murphy)

low: Reading my report on the state of minology in Russia which was accepted ry warmly and mentioned by the local ess. Following this there was a memorable ting, in the company of Bob Gimlin, John een and other friends, to the famous site the Patterson/Gimlin documentary film at uff Creek. It was my first and very enjoyle visit to the U.S. (Photo: D. Bayanov)

Left to right: Dr. Marie-Jeanne Koffmann, Dr. Grover Krantz, Dmitri Bayanov, in Moscow, 1997. The footprint casts are from the Koffmann Caucasus collection (almasty prints). Dr. Krantz passed away in 2002. He was one of the first prominent scientists to take an interest in the sasquatch/bigfoot issue. (Photo D. Bayanov.)

Author with Dr. Jane Goodall, Moscow, June 1999. Dr. Goodall has been interested in and supportive of our quest all along. What's more, she has publicly stated her belief in the reality of these enigmatic hairy hominids. Her wonderful work with known primate species is acknowledged and applauded world-wide. (Photo: D. Bayanov)

Author gifting his books to renowned geneticist professor Dr. Bryan Sykes, University of Oxford, after he delivered a lecture to members of the Smolin Hominology Seminar at the Darwin Museum in Moscow on August 21, 2013. I was happy to read in his e-mail to me: "I have benefitted greatly from reading your books which you kindly gave me..." (Photo: Alexei Mukhin)

small river named Voria which is near the author's country home. The water is fresh and clean enough for bathing. (Photo: Olesia Bayanova)

Author's country home; property shared with squirrels, hedgehogs, frogs, jays, tomtits, woodpeckers... and once visited by a moose. (Photo: Olesia Bayanova)

Author at a sandy hill by the river in August 2009. He and his family like to hike. (Photo: Olesia Bayanova)

Author's family portrait, taken on his property 30 miles north of Moscow, in August 2009. Left to right: son Ivan, granddaughters Olesia and Alina, wife Era, and author. (Photo: Olesia Bayanova)

View at sunset from author's apartment in Moscow of Novodevichy Convent, built in 1524, Moscow University (a

Daytime view of Novodevichy Convent and its pond, the place where the author exercises in the morning and takes walks (no longer jogs) around the pond inhabited by wild ducks. (Photo: Olesia Bayanova)

Author's earlier hominology books in English, French, German and Russian are shown below. The first published was *Wood Goblin Dubbed Monkey,* 1991 (sixth book shown). The first and last shown are essentially the same book (English), just different editions and publishers.

All of these books, and this current title, are the culmination of 50 years of research in hominology. They contain the record of scientifically unrecognized hominids throughout recorded history, both from the age of myth and folklore to the age of reality and the fringes of science. I believe we are on the threshold of resolving the issue and will hopefully write its final chapter. When we pass away, books remain and live on. (Photos: D. Bayanov)

CHAPTER 7

The Problem of Acknowledgement of Hominology by the Scientific Community

(Dmitri Bayanov's Address at the Meeting of Russian and Foreign Hominologists at the Darwin Museum, October 5, 2011.)

On behalf of Russian hominologists I greet our foreign friends and colleagues, as well as other guests of this conference. What are the meaning and goals of efforts exerted by hominologists? First of all in achieving recognition by world science of the existence of living relict hominids. Hominologists acknowledged the existence of these primates in the second half of last century and described their discoveries in many books and articles. But fundamentalists of academic science as a rule do not read our books and articles because they "know" that what is written in them just cannot be. They still believe that the Patterson/Gimlin authentic documentary is a fraud. What is the reason for this anti-scientific situation in world science? I will name three outstanding reasons:

Reason number one is that, due to its high volume of knowledge, world science has become highly specialized, while it is said that "a specialist is like a swollen cheek; his fullness is one-sided." Study of hominids different from us has long been conducted by paleoanthropologists who find and examine fossil material. The question of living non-*Homo sapiens* hominids is beyond their knowledge and horizon.

The scientific community believes the blind faith of these fine but narrow specialists that all such hominids died out at least tens of thousands of years ago. And this is not surprising, for evidence here is skeletal remains of long dead, not presently living specimens.

In the ancient world, and even later, when the number of sciences was smaller, when scientists were philosophers and encyclopedists, they well-knew of the existence of wild hairy bipeds whom they called troglodytes, that is, "cavemen." One of such philosophers and luminaries of natural history was Linnaeus the author of the terms *Homo sapiens* and *Homo troglodytes.* For the latter he also used the terms "*silvestris*" and "*nocturnes.*" Thus Linnaeus is the forefather of our direction of science.

Anthropologists are still unaware of the remarkable historical fact that the central and pretentious term of anthropology—*Homo sapiens*—appeared in science just in contrast to *Homo troglodytes*, the caveman, whose existence was known to naturalists of antiquity and the Middle Ages.

Reason number two is that the abundant biological information in folklore and mythology about these close-to-us bipedal primates is regarded by academic science as nothing more than popular fantasies based on superstition. In Russian folklore and demonology, the main term for these beings is "leshy" (forest man). The Nepalese word "yeti" used by our journalists instead of the Russian name "leshy" is a misnomer.

Reason number three is in the unique nature of relict hominids, in their special (including sexual) relations with people of our species, and in their extreme elusiveness and great parapsychological abilities.

Extreme secretiveness is also typical, as a rule, of people who establish and maintain friendly contact with these beings. In recent years however there have appeared several exceptions to this rule, which have brought most valuable results that will be reported I hope by my colleagues.

The reasons I named are closely interconnected and strengthen one another.

Modern world science is a conglomerate of numerous sciences, which lacked, however, a discipline devoted to the study of relict hominids, and thus their existence happened to be beyond the scope of the scientific community, and this despite the fact that the existence of bigfoot/sasquatch, for example, is well-known to the U.S. government.

The reasons why this knowledge is not becoming official and public are also well-known in America.

In 1999 I wrote a letter to Bill Clinton, then U.S. President, and sent him two of our books: *In the Footsteps of the Russian Snowman* and *America's Bigfoot: Fact, Not Fiction: U.S. Evidence Verified in Russia.* In the letter I asked him to pay attention to the bigfoot problem and defend the good names of Roger Patterson and Robert Gimlin who are constantly accused of having presented a fraud as a documentary film.

I received the following reply from the U.S. President: "Thank you for your kind gift and for sharing your thoughts and concerns. It's important for me to know your views. I'm glad you took the time to write." Alas, he did not do what I asked him for.

In short, relict hominids were unknown to science because there was no science to know them. Today we have such science. Since deep specialization of modern science is inevitable, the problem had to be resolved by creating one more special discipline—hominology, devoted to recognition and study of relict hominids.

The founding father of hominology in this country was Professor Boris Porshnev, historian and philosopher (life years 1905-1972). In my view, it is just because he was a humanist philosopher that the problem of man's origin excited him so strongly.

After half a century of our efforts under the hard conditions of self-funding and ostracism on the part of the scientific establishment, a few words on the situation right now.

Our North American colleagues have succeeded at last in taking clear videos of bigfoot/sasquatch, in addition to the Patterson-Gimlin documentary film. This happened thanks to the exceptions among bigfoot befrienders that I mentioned earlier. They have also obtained DNA samples of these hominids, with the analysis to be published in a scientific journal.

We wholeheartedly congratulate our North-American friends on these outstanding achievements. They give hope that the reality of relict hominids will soon be generally acknowledged.

Editors' Note: At press time, photographs, films/videos and DNA of bigfoot/sasquatch have not been verified by the general scientific community.

This will be the fuse of a scientific revolution in anthropology. The revolution in anthropology is about the most important question of human life; that is, about human nature, about knowing ourselves.

Besides, acknowledgement of relict hominids and hominology will raise a most important critical question of the quality and perspectives of modern world science, including the question of its lunatic militarization.

In conclusion we express heartfelt thanks to the Darwin Museum for actively supporting our activities from the very beginning in the middle of last century, for literally providing the roof over our heads.

We also extend our deep thanks to Aman Tuleyev, Governor of the Kemerovo Region, for his unexpected, very brave and many-sided support of our investigations at the present time. This is most fortunate and will be recorded in the history of science.

Thank you for listening.

Dmitri Yurevich Bayanov, Science Director
International Center of Hominology
Moscow, Russia

CHAPTER 8

Dr. Koffmann Replies to Professor Avdeyev

(Originally published in the April 1965 edition of *Nauka i Religia* [Science and Religion] magazine.
It is being provided here for a historical reference.)

Background

On October 1, 1964 the *Literaturnaya Gazeta* (Literary Gazette), published a letter from Professor Valeri Avdeyev under the heading, "Almastys—Where Are the Traces of Them?" It was this letter that prompted Dr. Koffmann's reply in April 1965 as indicated. Professor Avdeyev wrote::

> In popular-science literature from time to time the question is raised of the existence of hitherto unknown wild men. An article signed by four correspondents of the *TASS* news agency and *Nedelya* weekly has been published in Nedelya under the heading of "Do Almastys Exist?"
>
> They maintain that official science is wrong in rejecting the possibility of the existence of a wild man-like creature, unknown to science that is hiding in places difficult of access, because more and more testimony is being gathered to the effect that natives of the Caucasus have long known about him and some have seen him in our own times. Since the wild man, according to the article, is "nothing extraordinary" to the inhabitants of the Caucasus, the various nationalities there have different names for him. A list of these names is given in the article, and all of them mean man of the woods or wild man...
>
> It is perfectly clear that it will be possible to talk seriously of the existence of the almasty only after unquestionable material traces of their presence have been found.
>
> After my personal talks with her I sincerely

wished— and continue to wish—Jeanne Koffmann and her assistants the best of luck in their fascinating search. In the meantime, however, like many others, I do not believe in the existence of the almasty.

Reply from Dr. Marie-Jeanne Koffmann:

Esteemed Professor, I would like to comment on your views on the so-called snowman, which are shared by many others.

It is true that we do not yet possess serious material proofs that man-like creatures live in the Caucasus. The stories of eyewitnesses are as yet almost the only material we go by. This means that our claims are built entirely on their accounts. Do we have any reason to question them?

You, and those who think like you, solve that problem very simply. All reports concerning wild men, whether they originate from Tibet, the Pamirs, North America or the Caucasus, are dismissed as false. Discussion is thus closed.

But there is no unanimity among you, even in arriving at such an unsophisticated conclusion. Some of you think hundreds of eyewitnesses lie just like that, from a depravity inherent in human nature. Others see our informants as practical jokers, glad of a chance to put one over the scientists. Still others consider that all our witnesses are cowards who simply see things out of fright. Kinder critics are disposed to regard them as suffering from hallucinations. Finally, there are those who consider them to be backward ignoramuses, given to superstitious fears.

Two years ago, a well-known Moscow scientist, asked to at least look at my record of evidence, flatly refused to examine "old wives' tales gathered in the marketplace." Yet the file contained:

1) the record of a two- hour talk in one of the largest party district committees of Azerbaijan, signed by the second secretary of the committee, Dr. Kuliyeva, and a livestock specialist Akhadov;.

2) the report of militia captain Belalov;

3) an affidavit by Tairov, a research worker of the Academy of Sciences of Azerbaijan;

4) the testimony of Dyakov, an officer (Georgia);

5) the testimony of Shtymov, a Kabardian, dean of the faculty of pedagogics and psychology of the Pedagogical Institute in Kustanai;
6) the statement of Lt–Col. Karapetian of the Medical Corps, to the Academy of Sciences of the USSR;

7) the testimony of Kardanov, a Kabardian, a deputy to the Supreme Soviet of the USSR, etc.

The idea that the wild man is just a figure of folklore is ruled out by the testimony of witnesses not belonging to the local population. The view about ill-intentioned deception is incompatible with the testimony of persons who were unaware of the scientific controversy, and who enjoy considerable authority among the local inhabitants. The suggestion that it is a question of hallucination is refuted by the very nature of these hallucinations—the same in the case of hundreds of people at various times and in different localities.

I agree with you that if one wishes, one can collect any number of rumours about anything. However, people not only hear but they have the ability to evaluate what they hear. There is a method of ascertaining scientific truth by holding polls. A strictly worked-out system of compilation, analysis, comparative evaluations, verification and summarising of information guarantees a definite trustworthiness of the data received.

When I left for the Caucasus some years ago to verify the first reports that had reached us, I considered the possibility of wild men living there to be ridiculous, just as you and thousands of others do. It took a long time and hundreds of conversations before I reached the conclusion, and later the conviction, that I was dealing with realities.

You underline that you don't believe in the devil. I don't want to yield to you on this point, so I hasten to announce that I also don't believe in the devil. What is more, I don't believe in the almasty. I possess sufficient data to simply say that he exists!

CHAPTER 9

Brief Ecological Description of the Caucasus Relic Hominoid (Almasty)

Based on Oral Reports by Local Inhabitants and on Field Investigations
by
Marie-Jeanne Koffmann

Editors' Note: This article was translated from Russian into English by Dmitri Bayanov. It was originally published in *The Sasquatch and Other Unknown Hominoids* (1984), edited by: Vladimir Markotic´and Grover Krantz. It is being provided in this volume because of its great importance with regard to almasty research. All of the following is quoted material.

Introductory Remarks on this Paper: *This description of the natural history of the almasty of the Caucasus Mountain Region is a summary of much detailed and long investigations. Koffmann presents the material in a straightforward, factual manner that will leave many skeptics aghast. Koffmann gives a detailed account of what the almasty eats, when they can be seen, the territory they occupy and speculates on their possible population.*

Koffmann is convinced that these almasty are a drastically declining population. On the other hand, the North American sasquatch seems not to show any sign of a recent decline in population. (Editors: Vladimir Markotic´and Grover Krantz.)

The morphology of relic hominoids, though it varies to a certain extent from one geographic region to another, is so well-known today—both to the specialist and general public—that it seems preferable to dwell on some aspects of this species' biology.

My paper deals with the Caucasus and this should be specially stressed. The peculiarities of the Caucasus habitat have produced a unique situation for the hominoids and have deeply affected the ecology and ethnology of their local population.

Extending from the South-Russian steppes to the high plateaus of Anatolia in Armenia and Iran, the Caucasus takes up the whole isthmus between the Black and Caspian seas, covering an area of 440,000 km^2 (compared to Great Britain, 244,734 km^2 and Italy, 311,000 km^2).

The main geomorphological features of the region is the Greater Caucasus which crosses the isthmus with a barrier of parallel ranges attaining great heights and running uninterrupted for 1,200 km.

Despite the greatest variety of ethnic groups, tongues, religions and cultures, the people of Caucasus as a whole are characterized by the deeply antiquated nature of their customs and traditions. For thousands of years their livelihood has depended on the breeding of sheep and even today they retain the traits of ancient pastoral peoples, such as spiritual rectitude and simplicity, great hospitality, and keenness of observation.

In the Caucasus, a theater of the earliest civilizations, the hominoid population, pressed by *Homo sapiens* to lifeless uplands of rocky ranges, turned out to be surrounded and imprisoned by humans, as it were. The hominoid's resemblance to man aroused in the latter both fear of and pity for the creature. Not seeing special harm in them, but fearful of their great physical strength—and above all of their strange nature (neither man nor beast)—people preferred to maintain peaceful relations with the hominoids. The creatures even used to be offered food and old clothes by humans. Special sympathy used to be extended to their "women" with babies. The "almasty" (this is a Kabardinian name for the creatures which I am familiar with) have had enough presence of mind to profit fully by the proximity of man. The kind of relationship that exists between man and hominoid in the Caucasus is not to be met, as far as I know, anywhere else in the world at present.

The Caucasus habitat embraces practically all the territory of the isthmus: the almasty is not confined to a definite landscape or certain climatic, temperature or altitude conditions. The hominoid can be encountered in the rush flood-lands of the Podkumok and Terek Rivers, on the open and severe pre-Elbrus plateau, on the rock walls of the Greater and Lesser Caucasus, on the hot, dry plateaus of Karabakh and Armenia, in the dense, moist, subtropical forests of Kolkhida and the Talysh, as well as in the sand hills of the Caspian depression.

All these different landscapes serve as a natural background for the hominoid. With all that he tries to avoid open country, if possible, and clearly tends to inhabit woodlands. It is not fortuitous that in all Caucasian languages he is called "man of the woods," or "forest man."

Food and Eating Habits

As can be expected from the variety of his biotopes, the almasty is omnivorous. But one gets the impression that vegetarian food provides the staples of his diet, perhaps because of its abundance and easy availability. Coexistence with man has led to the almasty's partaking rather freely from the larder of his neighbor.

The following example from my field observations will help illuminate the almasty's alimentary ways:

A section of a corn field where an almasty "girl," sighted in the vicinity by the locals a short time before, must have been searching for sweet cobs, opening the wrapping leaves and taking a bite here and there, apparently to test the sweetness of the corn, without even tearing some of them off; this allowed us to obtain the creature's tooth line contours of the upper and lower mandibles; leftovers of a rat, having some characteristic peculiarities: the rat had been disemboweled very neatly and expertly, with the tail bitten off; fresh feces consisting almost exclusively of cherry stones, over 160 in all (cherries were not ripe at the time), and "tails," plus some seeds of different plants; a collection of almost fresh but unripe vegetables and fruits lying on a bedding of dry grass inside a low grotto rather difficult of access: the collection contained: eight potatoes, three apples, two small pumpkins, a half-nibbled corncob, a half-eaten sunflower center, some dog rose berries, plus four round pellets of horse dung (it is believed the almasty eats horse dung because of its salt content).

Among animal foods of the almasty, what strikes one as unusual is the placenta of domestic animals and, therefore, possibly of wild animals as well. The almasty's taste for it is so well-known that old herders, being in retirement and not quite realizing how different the conditions of keeping herds are at present, advised me to visit herds of horses and flocks of sheep in the spring to catch the almasty searching for placenta.

"You ask what [meat] the almasty eats? He eats placenta, he eats dead horses, and other dead animals" (Report No. 19 K). "Sheep were giving birth then, and the almasty was taking their placenta. Once, when I came nearer, he grabbed the placenta and, grumbling, went away behind the stones" (Report No. 111 K).

Wild Plants	Cultivated Plants
All kinds of Caucasian wild fruits and berries	All kinds of fruits
	Watermelon
Sorrel	Pumpkin
Bugloss	Tomato
Wild chervil	Onion
Cow-parsnip	Green Pepper
Shepherd's purse	Potato
Meadow-rue	Corn
Dog rose	Sunflower
Ashberry	Hemp
Moss	
Ecphymas, fungus formation on trees	**Foods Taken from Man**
	Milk
Water mould	"Airan" (sour, fermented milk)
	Cheese
Animal Foods	Bread
Carrion	Flour
Placenta of ungulates	Eggs
Frogs	Meat
Frog's eggs	Honey
Lizards	Cooked meals (soup, porridge)
Tortoises	
Mice	**Mineral Food**
Squirrels	Rock-salt
Rats	Mineral concentrations at mineral water springs
Bats	
	White clay

The following lists include only those foods that my informants insisted they had actually observed the almasty eating:

It is hard to determine the proportion of human-type food in the almasty's diet, but I think it is quite extensive.

Information on the hunting activities of the hominoid in the

Caucasus is very scanty in comparison with some regions of Eurasia and America. The almasty's low hunting activity can be explained by good vegetarian feeding grounds in the Caucasus. It is worth mentioning that, according to locals, the almasty can subsist on very little food, but when he gets to eating he does so in a greedy and rapid manner, yet never for a second letting down his guard.

As for drinking, the almasty prefers spring water. "Having reached the spring, he knelt down, placed his hands on the ground and, just like man, bent to the water and began to drink. He was 15m from me. He drank for a long time, taking short intervals: after drinking for a while he would raise his head, glancing this way and that, and then drink again. He drinks like a horse, sucking in the water through his pressed lips" (Report No. 54G). Incidentally, the chimpanzee also drinks through his pressed lips.

The almasty is believed to be able to do without drinking for long periods. When feeding in a corn field he can stay put for several days, content with the liquid contained in the food.

Feeding mainly on plants, the almasty is bound to be dependent on the vegetative conditions of his feeding grounds and to change them according to the season. And this is just the case: the annual cycle of migration is very well-defined. In its simplest form this cycle is manifest; for example in Northern Azerbaijan, encounters are registered exclusively in the summer and fall months, i.e., the season of chestnuts, acorns, walnuts, hazelnuts, and all wild and cultivated fruits.

In the Northern Central Caucasus (Kabardino-Balkaria), roughly between Pyatigorsk and Nalchik where the difference in terrain elevation is not as abrupt as on the southern slopes, and where the vegetation zones change rather smoothly as a consequence, seasonal migration is not as sharply defined but is nevertheless apparent.

Sightings

On the average in the Caucasus sightings are distributed according to season (see Figure 1). Although I did not copy this graph from John Green's *The Sasquatch File,* p. 63, I guess it will give him and our other American colleagues as much pleasure as his graphs gave to us.

Some regions, for example the Central Caucasus, are more

important than others because encounters are taking place here almost throughout the year. However, here as well they are more numerous in summer (both the almasty and the local residents are more active then), and rare in early spring (in March-April, just as they are in the western mountainous regions of Canada and the USA). The drop in the spring cannot be explained by a lesser activity of the people. On the contrary, the population is very active in the spring: a time of lambing, of sheep shearing, of spring agricultural work. But for a herbivore it is the hardest time of the year, worse even then the winter which is usually rather mild in the Northern Caucasus and quite temperate in Transcaucasia.

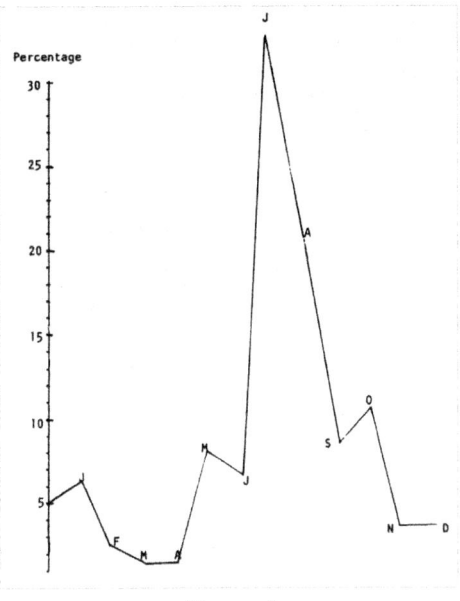

Figure 1
Average Monthly Encounters in the Caucasus.

I have the impression that for this period the almasty abandons the empty fields of the foothills, where people dwell, and moves higher into the forest zone. There he can still find dry wild fruits, left from the previous year, and a lot of roots on the southern slopes where in some valleys it is quite warm.

I also think that during the leanest season the almasty can fall to "sleep." This is not true hibernation, but a kind of protective reaction against adverse conditions, accompanied by a state of low metabolism, which helps the creature weather the adversity. Maybe this is why the Kabardinians, when speaking of something only superficially pleasant, say "There is as much fun in it as in the sleep of an almasty."

I've heard people say that the creature must store up food for the winter, but those were just suppositions because nobody had ever actually found such stores.

We are also well aware of the daily activity rhythm of the

almasty which transpires from the answers to two questions: at what hour of the day or night did the encounter take place? And what was the almasty doing when encountered? During the night (33% of encounters) the creature is busy looking for food; that is, to put it bluntly, engaged in thieving when in the area of man's habitation. At dawn (11% of encounters) the night activity ends. The almasty sleeps a lot in the daytime (33% of encounters), this time very soundly, sometimes unaware of people approaching. "We followed the trail which led into a spot overgrown with tall weeds. He was there, lying on his back and sleeping peacefully. There were eight of us; we surrounded him and stood looking" (Report No. 126 K). "I was walking in the grass and almost stepped on a sleeping almasty!" (Report No. 135 K). Thus it follows that the almasty is a crepusculo-nocturnal creature. Can it be accidental that Linnaeus attached three epithets to his second species of *Homo:* troglodytes, nocturnus, and sylvestris? Only the third is applicable to an ape, whereas all three fit the relic hominoid.

We know nothing of any weekly or monthly cycles of almasty activity except for the fact that he is always on the move. Migration seems to be a specific characteristic and is an outstanding peculiarity of all populations of hominoids in different habitats the world over. Like his counterparts, the almasty does not use a shelter for a long time which, incidentally, testifies against the hypothesis of winter stores. In the winter he rests in chance refuges (an abandoned herder's cabin, a haystack, a grotto), in the summer simply lying on the ground in tall weeds or climbing a tree. On the ground he makes a lair with a bedding of rags and soft grass. He ties up the tops of tall weeds (making knots is one of his favorite pastimes) and covers this frame with a canopy of burdock leaves. Nests in trees are made of big branches, which are broken, intertwined and matted into a soft bedding.

Territories

It is very hard to get an idea of the size of an area on which a specimen or a group is wandering. It is not even known if there are any dividing lines between such territories within a geographical region. However, our preliminary data are in agreement with Dr. Krantz's surmising as regards territorial distribution of individuals.

With the help of local witnesses, I managed to "spy" on the presence of five specimens, each in his own area, on a territory of about 250 km^2 during two summer seasons, Unfortunately, it was then that I had to interrupt my investigations for a number of years. Returning six years later I managed to recognize two of the old-timers by their behavior and territories. In his temporary territory each individual was acting covertly and constantly moving from place to place, so that the information which reached us about his activities was invariably late by two or three days. The territories of the individuals also overlapped.

Population

This brings me to one of the central questions which is hard to by-pass and which is difficult to answer—the number of hominoids in the Caucasus. I am only certain of one thing: the number is falling catastrophically and must presently be at a critical or even lower level, a process which began about 40 or 50 years ago and has been gaining momentum. In the 16 years of my work in the Central Caucasus the number of encounters has dropped significantly. There are a number of factors to explain the sorry plight of the hominoid in the Caucasus: war operations in the recent past, a demographic explosion, fundamental anthropogenic transformations of landscapes, a sudden advent of plenty of modern machinery in the recently feudal rural areas, etc.

If I do not risk pronouncing an absolute figure for the Caucasus population of the hominoid, then at least I can say something about its sex and age composition in the recent past. It is characterized by stability in each area and changes from area to area. Thus, in Northern Azerbaijan males accounted for 60% of identified individuals (in many cases the sex of the encountered hominoid remains unidentified), females accounted for 40%, while the young in that region in the last decades are not registered in my files at all.

Quite a different situation is seen in the north of the Central Caucasus: 54% females, of which 21% are with young, 18% juveniles (8–16 years old?), only 14% males, and 14% of those sighted in groups. (The data of 1930-1965).

The groups of almasty, invariably including the young, consist of four to six and, in rare cases, up to ten. In the old days such groups were not fearful of the local inhabitants and were noted for their boisterous behavior: it appeared that the almasty would often start fights among themselves to the accompaniment of cries, screams and "weeping," which would rather suddenly change to peaceful mumbling in the intervals between fights.

Miscellaneous Gatherings

Not discussing the great variety of the creature's vocalization, I want only to note that the almasty is given to mumbling even when alone. He is also capable of a cry of tremendous power, which is heard far and wide in the night and probably serves as a call.

His eyesight is excellent, both day and night vision, the latter confirmed by his "eye-shine," noted by almost every nocturnal observer of the almasty (Napier 1973:169).

There is but little and contradictory information on the creature's sense of smell. It seems that the peculiarly nasty smell reported of some specimens could have a communicative significance. It is of such strength that even man can identify it from afar. Every time my informants mentioned this smell ("stinking like a toilet," "like a dead dog"), it was a case of a male or of a specimen of unidentified sex, which, by certain traits, appeared to be male. I have no information on identified almasty females giving off this smell.

I have two descriptions of almasty births, one of them of twins, "The newborns are just like human, but small, no more than two kilos each (4.4 pounds), but otherwise just like human, you wouldn't tell them apart. The skin was pink...without hair" (Report No. 85 K). According to my informants, a one-year-old almasty is already covered with short hair.

The age of young specimens is judged by witnesses using human standards and, naturally, considerable mistakes are thus possible. The "youngsters," beginning to appear alone, are guessed to be "eight years' old," and this may indicate that the young almasty require extended care by the mother.

In the unanimous opinion of old men among locals, the life span of an almasty is long and comparable to that of man. They cited examples of long-term associations of humans, including the parents and grandparents of the informant, with a particular almasty, the latter receiving food offerings from the family over many years.

The almasty has no natural enemies (I am not taking into consideration the rare brawls with a bear when feeding on wild raspberries or currants), and is hardly prone to traumatism. He is only threatened by wolves and the hefty Caucasian shepherd dogs which can induce panic in young and female almasty. (I have descriptions of almasty corpses presumably bitten to death and disfigured by dogs.) But as a rule, the almasty probably dies a slow, natural death (of old age or some ailment), which gives him time to seek a covert refuge.

The mathematical and graphical analysis of sighting reports and their interpretation in terms of anatomy have previously revealed the following:

1. The trustworthiness of morphological elements considered separately (the low and receding forehead with prominently protruding eyebrows, platyrrhini, the chinless mandible, etc.).

2. The presence of classical architectonic correlations between these elements.

3. The coherence of morphology and function (for example, lack of thenar, and way of grasping). (Koffmann 1966, 1967a).

All categories of information about the almasty—ecological, morphological, and ethnological (the last two left out here)—converge to produce a coherent and viable image; that of a primate, a hominoid—anatomically sound, biologically plausible, anthropologically sensible.

Being late by 30–40 years, we witness the end of the Caucasus hominoid. Remnants of the recently-numerous population consist of only separate individuals roaming in solitude among the fragments of their habitat.

That is why, on the one hand, we are so well informed about the outward appearance of the almasty and the main aspects of his way of life. This information has been supplied mainly by people of the older generation who are dwindling themselves. And that is why, on the other hand, we are so hard put to produce concrete proof of almasty's existence.

References cited for this paper:

Green, John (1973). *The Sasquatch File.* Agassiz: Cheam Publishing.
Koffmann, M. J. (1966). Predvaritel'nye itogi izucheniia relik tovykh gominoidov na Kavkaze (Preliminary results of relic hominoid research in the Caucasus) Session of the Geographic Society of the USSR Academy of Sciences, March 22.
Koffmann, M. J. (1967). "Summary of report at the 1st Conference on the Problems of Medical Geography of Northern Caucasus" (in Russian), Leningrad.
Koffmann, M. J. (1967). "Report à Museum d'Histoire Naturelie," Paris (Report to the Museum of Natural History, Paris).
Krantz, Dr. Grover S. (1983). "Research on Unknown Hominoids in North America" from Vladimir Markotic´, (editor) and Grover Krantz, (associated editor) (1984). *The Sasquatch and other Unknown Hominoids.* Calgary, Alberta: Western Publishers.
Napier, John (1973). *Bigfoot.* New York: Dutton Publishing.

Author's Afterword: As stated, this article was written by Dr. Marie-Jeanne Koffmann. Historically, Dr. Koffmann is, after Professor Porshnev, the main contributor to Russian hominology. It was in her expeditions that I and many others were introduced to "practical" hominology. Professor Porshnev started Russian hominology in theory; Dr. Koffmann did it in practice.

We owe a great debt to Dr. Koffmann for her remarkable work. I greatly value my close association with her for nearly 50 years.

Dr. Koffmann in earlier years. She is now in her mid 90s and resides in a rest home. (Photo: Author's file)

Final Concluding Remarks

Well, dear reader, here you have my most important articles and papers on the past and present evidence of homin existence, and my plea to have hominology recognized as a valid scientific discipline so as to gain the involvement of more scientists and resources to resolve the issue.

As testified by this book, I have devoted 50 years to this subject in close cooperation with friends and colleagues in North America. This convinced me that there does exist on our planet a kind (or kinds) of human beings different in appearance and way of life from all known types of our own species. With time pressing on me, all I can now do is pass the baton to the next generation of dedicated truth seekers.

The "re-discovery" of most secretive and elusive, as well as ubiquitous "forest people" will provide many missing clues to human evolution and open the door to a much better understanding of who we are. I hope that this will bring about much greater cooperation in the world which is essential for the survival of civilization.

Thank you for going on this journey with me.

Dmitri Bayanov
Moscow, Russia
Spring, 2015

BIBLIOGRAPHY

Note: Secondary Bibliographical references for Chapter 2 are cited in the Original Source section under each entry.

Bayanov, Dmitri (1991). *Wood Goblin Dubbed Monkey* (in Russian). Moscow: Obshestvo po izucheniyu tain i zagadok zemli.

Bayanov, Dmitri (1996). *In the Footsteps of the Russian Snowman*. Moscow, Russia: Crypto-Logos Publishers.

Bayanov, Dmitri (1997). *America's Bigfoot: Fact, Not Fiction--U.S. Evidence Verified in Russia*. Moscow, Russia: Crypto-Logos Publishers.

Bayanov, Dmitri (2001). *Bigfoot: To Kill or to Film-The Problem of Proof.* Burnaby, BC: Pyramid Publications.

Bayanov, Dmitri (2002). "Some Thoughts on the Origin of Speech" (article).

Bayanov, Dmitri (2011). *Bigfoot Research: The Russian Vision*. Surrey, BC: Hancock House Publishers.
Moscow, Russia: Crypto-Logos Publishers (2007).

Bayanov, Dmitriand Igor Burtsev (1974). "Reply" (to Comments). *Current Anthropology,* 15(4):452–456.

Bayanov, Dmitri and Igor, Burtsev (1976). On Neanderthal vs. Paranthropus. Current Anthropology, 17(2):312-318.

Bernheimer, R. (1952). *Wild Men in the Middle Ages.* Cambridge: Harvard University Press.

Bindernagel, John (2010). *The Discovery of the Sasquatch.* Courtenay, BC: Beachcomber Books.

Blumenbach, Johann (1775). *Manual of Natural History.*

Bondt, Jacob (also Bontius, Jacobus) (1592-1631), *Historiae naturalis et medicae Indiae orientalis.*

Bontii, Iacobi (1658). *Historiae Naturalis et Medicae Indiae Orientalis.*

Burns, John W. (1940). "The 'Hairy Giants' of British Columbia," *The Wide World* magazine, January 1940, Vol. 84, No. 502.

Coleman, Loren (2003). *Bigfoot! The True Story of Apes in America*. New York, NY: Paraview Pocket Books.

Daegling, David (2004). *Bigfoot Exposed: An Anthropologist Examines America's Enduring Legend.* Walnut Creek, California: AltaMira Press.

Dahl, Vladimir (1880 first edition). *Dictionary of the Russian Language.* Russia: (many editions).

Darwin, Charles (1860 first edition). *A Naturalist's Voyage Round the World.*
Darwin, Charles (1929). *The Origin of Species by Natural Selection.*
Davitashvili, L. S. (1948). *History of Evolutionary Paleontology from Darwin to Our Days.*
Diodorus, Siculus (1774). *Bibliotheca Historica* (translated from Greek into Russian). St. Petersburg. Part I, p.141.
Genus (1962). Vol.18, No.1-4, p.4. Rome, Italy.
Goodall, Jane (1996). *My Life with the Chimpanzees.* Aladdin (publisher).
Green, John (1973). *The Sasquatch File.* Agassiz, BC: Cheam Publishing.
Green, John (1978). *Sasquatch: The Apes Among Us.* Surrey, BC: Hancock House Publishers.
Green, Mary A., and Janice Carter Coy (2002). *50 Years with Bigfoot.* Cookeville, Tennessee:. Green and Coy Enterprises.
Healy, Tony, and Paul Cropper, (2006). *The Yowie: In Search of Australia's Bigfoot.* Sydney, Australia: Strange Nation.
Hesler, Heinrich von (1400s). *Apocalypse.*
Heuvelmans, Bernard and Boris Porshnev, (1974). *L'Homme de Neanderthal est toujours vivant.* Plon, France: Librairie Plon.
Huxley, Thomas (1863). *Man's Place in Nature.*
Koffmann, M. J. (1966). "Predvaritel'nye itogi izucheniia relik tovykh gominoidov na Kavkaze ("Preliminary results of relic hominoid research in the Caucasus"). Session of the Geographic Society of the USSR Academy of Sciences, March 22.
Koffmann, M. J. (1967). "Summary of report at the 1st Conference on the Problems of Medical Geography of Northern Caucasus (in Russian): Leningrad.
Koffmann, M.J. (1967). "Report à Museum d'Histoire Naturelle, Paris." (Report to the Museum of Natural History, Paris).
Korogly, K. (1983). "Interactions of eposes of the peoples of Central Asia, Iran, and Azerbaijan" (in Russian). Moscow: Nauka.
Krantz, Grover (1980). "Sapienization and Speech." *Current Anthropology,* 21(6):773–792.
Krantz, Grover (1983). "Research on Unknown Hominoids in North America," from Vladimir Markotic´,(editor) and Grover Krantz (associated editor) (1984). *The Sasquatch and other Unknown Hominoids.* Calgary, Alberta: Western Publishers.

Kuhn, Thomas S. (1962). *The Structure of Scientific Revolutions.* Chicago: University of Chicago Press.

Linnaeus, C. (1758). *Systema Naturae.* Stockholm: Laurentii Salvii Holmiae.

Linnaeus, C. (1760). "VI – Anthropomorpha." Amoenitates Academiae. Literaturnaya Gazeta (Literary Gazette), Russia.

Lucretius T. (1947). *De rerum natura.* London: Oxford University Press.

Markotic´, Vladimir (editor) and Grover Krantz (associated editor) (1984). T*he Sasquatch and other Unknown Hominoids.* Calgary, Alberta: Western Publishers.

Masudi, A. (1841). *Ali Masudi's Historical Encyclopedia.* London: Spenger.

Moskowitz Strain, Kathy (2008). *Giants, Cannibals & Monsters: Bigfoot in Native Culture.* Surrey, BC: Hancock House Publishers.

Murphy, Christopher L. (2004). *Meet the Sasquatch.* Surrey, BC: Hancock House Publishers, pp. 31, 33.

Murphy, Christopher L. (2010). *Know the Sasquatch: Sequel and Update to Meet the Sasquatch.* Surrey, BC: Hancock House Publishers.

Napier, John (1973). *Bigfoot.* New York: Dutton Publishing.

Nizami, A. (1963). *Chahar maqala* (translated into Russian). Moscow: Eastern Literature Publishing House.

Pausanias (1913). *Pausanias's Description of Greece.* London: MacMillan.

Pliny (1979). *Pliny the Elder's Natural History.* Cambridge: Harvard University Press.

Plutarch (1792). *Plutarch's Lives.* London: Dilly.

Porshnev, Boris (1963). *The Present State of the Question of Relict Hominoids* (in Russian). Moscow: VINITI.

Porshnev, Boris (1974). "The Troglodytidae and the Hominidae in the Taxonomy and Evolution of Higher Primates." *Current Anthropology,* 15(4): 449-450.

Reder, D. (1965). *Myths and Legends of the Ancient Land of the Two Rivers* (in Russian). Moscow: Nauka.

Roosevelt, Theodore (1893). *The Wilderness Hunter – Outdoor Pastimes of an American Hunter.* New York, NY: G.P. Putnam's Sons.

Sanderson, Ivan T. (1961). *Abominable Snowmen: Legend Come to Life.* Kempton, Illinois: Adventures Unlimited Press (2006 edition), p. 10.

Sandgathe, D. M. with H. L. Dibble, P. Goldberg, S. P. McPherron, A. Turq, L. Niven and J. Hodgkins (2011). "On the Role of Fire in Neandertal Adaptations in Western Europe." Evidence from Pech de l'Azé IV and Roc de Marsal (excavation sites), France: *Paleoanthropology,* 2011:216-242.

Schiltberger, J. (1879). *The Bondage and Travels of Johann Schiltberger.* London.

Shackley, M. (1983). *Still Living? Yeti, Sasquatch and the Neanderthal Enigma.* New York: Thames and Hudson.

Shein, P. (1893). *Materials for the study of everyday life and language of the Russian population of the North-Western Region* (in Russian). St. Petersburg. Vol. 2, p. 526.

Singh, J. and R. Zingg (1942). *Wolf-children and Feral Man.* New York and London: Harper.

Sinistrari, L. (1875). "De la demonialite et des animaux incubes et succubes..." Translated from Latin into French by Isidore Liseux. Paris.

Sprague, Roderick and Grover Krantz – editors (1977). *The Scientist Looks at the Sasquatch.* Moscow, Idaho: University Press of Idaho (A division of the Idaho Research Foundation), pp. 57, 58.

Strabo (1964). *Geographica* (translated from Greek into Russian). Leningrad: Nauka. p. 449.

Tyson, Edward (1699). *Ourang-outang, sive Homo sylvestris* (Anatomy of a Pygmie Compared with that of a Monkey, an Ape, and a Man).

Wagner, M. (1796). *Beitrage zur philosophischen Anthropologie.* Vienna.

Zelenin, D. K. (1916). *Essays on Russian Mythology.* Petrograd, Russia.

———— Geographic Society of the USSR Academy of Sciences, March 22.

———— *Everyman's Encyclopedia* (1913).

———— *Questions of Philosophy* (Russian Journal). Issued by the Institute of Philosophy under the auspices of the Soviet Academy of Sciences.

General Index

Abkhazia (territory; eastern coast of the Black Sea), 31
Adam (Biblical first man), 24, 111, 133, 147
Agogino, George, Dr., 12
Akhadov, Mr. (livestock specialist), 182
al-'Arudi, Nizami, 15, 36. 61
Aristarchus, (Greek astronomer and mathematician), 144
Athabaskan (people), 105
Austin, Ron, 62
Avdeyev, Valen, Prof., 9, 181
Babylon, 32
Babylonian epic, 24
Bashkir folklore, 67
Bayanov, Alina, 173
Bayanov, Dmitri (Frontispiece), 2
Bayanov, Era, 173
Bayanov, Ivan, 173
Bayanov, Olesia, 173
Beaverton, Oregon, 47
Beeson, Richard, 46, 47
Belalov, (militia captain), 182
Bernheimer, Richard, 16, 84, 111, 116, 124
Bible, The, 22, 31, 32, 33, 133
Bielorussia, 24, 29
Bielorussian folklore, 36
Bielorussian songs, 28
Bigfoot Symposium, Willow Creek, California, 41, 167
Bilibin, Ivan, 25
Bindernagel, John, Dr., 143, 145
Blackie (bigfoot name), 97
Blue Mountains, Washington, 74
Blumenbach, Johan, 21, 133

Bontius, Jacobus, 19, 130, 131, 132, 133, 135, 136
Bourtsev, Igor (see Burtsev)
Bourtseva, Alexandra, 162
Bourtseva, Lydia, 163
Brasov (Romania)), 18
Bruno, Giordano, 157
Bureau of American Ethnology, 99
Burenhult, Niclas, 35, 36
Burns, J. W., 30, 62, 65, 66, 85
Burns, M., 12
Burtsev, Igor (also Bourtsev), 11, 107
Bury, Howard, Colonel, 81
Butler, James, 98
Bykova, Maya, 82
Byrne, Peter, 12
Carter Coy, Janice, 41, 56, 66, 73, 96, 102, 107, 111, 112, 115, 117, 119, 120
Carter farm, 68, 119
Carter, Robert, Sr., 117, 119
Cheeco (bigfoot name), 101
Chehalis (people), 62, 65
Cherokee (people), 97
Chickasaw (people), 97
Chief Splashing Water, 71
Chinook (people), 69
Choctaw (people), 99, 100
Christ, Jesus, 34
Christianity, 23, 34, 88, 116, 161
Chusovaya River, Urals, 166
Chuvash (people), 107
Chuvash folklore, 6, 30, 61, 62
Clackamas River, Oregon, 69
Clinton, Bill, (President), 178
Coeur d' Alene (people), 78

Columbia River,
  Washington/Oregon, 92
Comanche (people), 52
Copernicus, Nicolaus, 144, 156
Cropper, Paul, 119, 120
Dahinden, René, 12, 65, 154
Dahl, Vladimir, 29, 39
Dalles, Oregon, 92
Dart, Raymond A., Dr., 12
Darwin, Charles, 21, 135, 141,
  147, 157
Davis, Don, 68
Dawkins, Richard, 157
de Bondt, Jacob, 130
Disotell, Todd, 140
Drava River, Europe, 117
Duncan, Will, 127
Dyakov, Mr. (officer – Georgia),
  182
Ebu Gogo myths, 138
Enkidu (homin), 32, 33
Eskimaun (people), 106
Eve (Biblical first woman), 24
Fahrenbach, Henner, Dr., 38, 82
Forth, Gregory, 138
Fox (bigfoot name), 96, 117, 119
Fraser River, B.C., 62
Fusch, Ed, Dr., 68
Galileo, 156, 157
Gee, Henry, 139
Gilgamesh, 24
Gimlin, Robert (Bob), 167, 178
Gini, Corrado, Dr., 11. 12
Goodall, Jane, Dr., 155, 156, 169
Gorbachev, Mikhail, 150
Gould, Stephen J., 21, 133
Green, John, 12, 30, 65, 66, 85,
  106, 154, 167, 188,
Green, Mary, 41, 111

Grinyova, Nina, 165
Haas, George. 154, 155, 156, 159
Haldane, John, 159
Hare (people), 105
Harrison Mills, B.C., 62
Healy, Tony, 119, 129
Hebrew folklore, 32
Hesiod (Greek poet), 37
Heuvelmans, Bernard, Dr., 10. 11,
  12, 133, 135,
Hill, Osman, W.C., Dr., 12
Hobbit (early human), 114
Hoppius, Christian, 20, 130, 148
Humbaba (demon), 32
Huxley, Thomas H., 21, 43, 109,
  133, 148
Inquisition, The, 34
International Committee for the
  Study of Hairy Hominoids, 11
Ishtar (goddess of love), 32
Islam, 161
Isles of the Satyrs, 14
Izzard, Ralph, 12
Judaism, 161
Kabardians (people), 30
Karapetian, Dr. and Lt. Colonel,
  163, 183
Kardanov, (Government official;
  deputy to the supreme Soviet of
  the USSR), 183
Katayev, Alexander, 166
Kazakh folklore, 36, 67
Keller Butte, 90
Keller, Washington, 89
Khakhlov, Vitaly, Prof., 148
Khan of Siberia, 15
King Gilgamesh, 32, 33
King Midas, 31

King Numa Pompilius, 31
Kinion, Robert, 127
Klickitat (people), 83
Knatterud, Erik, 35
Knights, Roger, 158
Koffmann, Marie-Jeanne, Dr., 2, 7, 9, 60, 129, 168, 181, 182, 184, 194
Komi (people), 26
Kootenai (people), 78
Koran, 67
Krantz, Grover, Dr., 41, 46, 116, 168, 190
Kronstadt (Romania), 18
Kuhn, Thomas, 144, 145, 158
Kuiliyeva, Dr., 182
Kutlukai (legendary hunter), 67
Kwakiutl (people), 69
Lake Pairon, Tajikistan, 165
Linnaeus, Carl, 10, 19, 21, 35, 130, 131, 132, 133 142, 147, 148, 177
Long, Serephine, 62, 63, 65, 66
Lovrencevic, Zvonko, 116
Lucretius, Carus, 10,
Lupercalia (celebration), 27
Magnus, Albertus, 16
Magnus, Olaus, 35
Makarov, Vadim, 164, 166
Mansi (people), 25
Markotic´, Vladimir, 116
Mashkovtsev, Alexander, 2
Masudi, Abul Hasan Ali, 15
Menlung Glacier, Himalayas, 113
Milic, Goran, 113
Modoc (people), 79
Montagna, William, Dr., 47
Mordova (people), 27
Morley, John, 114, 125

Moscow University, 174
Moscow Zoological Museum, 19
Moses (Biblical), 33, 34
Moskowitz Strain, Kathy, 8, 47, 48, 49, 57, 68, 107, 108, 109, 111, 112
Mosopelea (people), 55
Murphy, Christopher, 48, 65, 113, 106, 163, 167
Muskogean (people), 99, 102
Nage (people), 138
Napier, John, Dr., 12, 21, 145, 146
Nehalem (people), 75
Nehalem River, Oregon, 76
Nelchina (people), 106
Nelson, Scott, 66, 115
New York University, 140
Nisqually (people), 83
Nogai folklore, 67
Nogai nobility, 67
Notre Dame, France, 16
Nymphaion (Greek colony), 15
Ojibwa (people) 57
Orthodox Church, 27
Ostman, Albert, 30, 38, 44, 65, 66
Pan (mythology), 38,114
Patterson, Roger, 178
Patterson/Gimlin documentary film, 47, 68, 144, 146 162, 163, 167, 177, 179
Pausanias (Greek traveler and geographer), 14, 35
Pedagogical Institute, Kustanai, 182
Penutian (people), 75, 91
Pliny the Elder, 14, 19, 37, 136
Plutarch, (Greek historian), 114
Podkumok River, Caucasus, 185
Pope Innocent VIII, 34

Pope John Paul II, 157
Porshnev, Boris, Prof., 2, 7, 8, 9, 10, 11, 12, 21, 40, 43, 68, 108, 111, 133, 145, 146, 148, 149, 151, 154, 179
Portland State University, 45
Puskin Fine Arts Museum, Moscow, 114
Puyallup (people), 83
Pye, Lloyd, 113
Regional Primate Research Center, Beaverton, Oregon, 47
Relict Hominoid Inquiry, 12
Rinchen, P. R., Dr., 12
Roosevelt, Theodore, 22
Rumphius, Georg, 19
Russian State Library, Moscow, 135
Saint Savvati, 161
Saint Zosima, 161
Salishan (people), 87, 89
Samos (Greek island), 144
Sanderson, Ivan, 10, 11, 12, 129, 154
Schiltberger, Johann, 15
Schliemann, Heinrich, 44
Semur en Auzois, Burgandy, France, 16
Sevastianov, Alexander, 130, 131, 135
Shaitan-Kudey clan, 67
Shasta (people), 49
Shawnee (people), 58
Sheba (bigfoot name), 96, 101
Shipton, Eric, 113
Short, Bobbie, 68,
Shoshone (people), 51
Shtymov (dean), 182
Siculus, Diodorus, 33

Silenus (mythology), 14, 31,114
Siminole (people), 102
Sinistran, Luigi Maria, 34, 35
Sioux (people), 56
Slick, Tom, 12
Smoky Mountains, Tennessee, 42
Smolin, Pyotr, 2
Solo (name of a man), 121
Spirit Lake, Washington, 83
Spokane (people), 87
Sprague, Roderick, 46
St. Augustine, 34
St. Petersburg, Russia, 20, 21
Stalin, Joseph, 174
Steenburg, Thomas, 106
Strasenburgh, Gordon, 154
Sulla Lucius Cornelius (Roman general), 15, 114, 115
Suttles, Wayne, 45, 46, 66
Swift, Jonathan, 56
Sykes, Bryan, Dr., 170
Tairov, Mr. (researcher), 182
Talmud, 39
Tatars (people), 30, 31
Tattersall, Dr., 140
Temple of Silenus, 31
Tench, C. V., 65
Terek River, Caucasus, 185
Tillamook Bay, Oregon, 75
Titmus, Bob, 12
Tobias, Philip V., Prof., 12
Trachtengerts, Michael, Dr., 72, 74, 107
Troy (city), 44
Tsimshian (people), 108
Tuleyev, Aman, 180
Tyson, Edward, 132
Ukranian songs, 28
University of Idaho, 46

van den Bergh, Gerd, 139
van Heerwarden, Mr., 131
Vereshchagin, Nikolai, Prof., 8, 149
Vladimir Province, 68,
von Hesler, Heinrich, 111
Voria River, Russia, 171
Wagner, Michael, 18
Walla Walla, Washington, 75
Wenatchee (people), 89
Wenatchee Lake, Washington, 89
Wenatchee River, Washington, 89
Wheatcroft, Wilson, Dr., 128, 130, 134, 135
Willow Creek Bigfoot Symposium, 41, 167
Wintu (people), 50
Yakima (people), 83, 91, 96, 97
Yakima County, 92
Yakima River, Washington, 92
Zakirov (Aslon), 90
Zelenin, D. K., 68
Zoroastrianism, 23
Zrinski, Peter, 118

other cryptozoology titles from HANCOCK HOUSE PUBLISHERS

Sasquatch in BC
Chris Murphy
978-0-88839-721-8
5½ x 8½, sc, 528pp
$29.95

Raincoast Sasquatch
Robert Alley
978-0-88839-143-8
5½ x 8½, sc, 360pp
$29.95

Giants, Cannibals & Monsters
Kathy Strain
978-0-88839-650-1
8½ x 11, sc, 288pp
$39.95

Sasquatch in Alberta
Thomas Steenburg
978-0-88839-408-8
5½ x 8½, sc, 116 pp
$19.95

Bigfoot Film Controversy
Chris Murphy
978-0-88839-581-8
5½ x 8½, sc, 240pp
$22.95

Bigfoot Research
Dmitri Bayanov
978-0-88839-706-5
5½ x 8½ sc, 424pp
$29.95

Bigfoot Film Journal
Chris Murphy
978-0-88839-658-7
5½ x 8½ sc, 106pp
$29.95

The Asian Wildman
Jean-Paul Debenat
978-0-88839-719-5
5½ x 8½ sc, 176pp
$17.95

Sasquatch Bigfoot
Thomas Steenburg
978-0-88839-685-3
5½ x 8½, sc,128pp
$12.95

Sasquatch: *mystery of the wildman*
Jean-Paul Debenat
978-0-88839-685-3
5½ x 8½ sc, 428 pp
$29.95

Best of Sasquatch Bigfoot
John Green
978-0-88839-546-7
8½ x 11, sc, 144pp
$19.95

Russian Hominology
Dmitri Bayanov
978-0-88839-736-2
5½ x 8½, sc, 204 pp
$19.95

Know the Sasquatch
Chris Murphy
978-0-88839-657-0
8½ x 11, sc, 64 pp
$34.95

Hoopa Project
David Paulides
978-0-88839-015-8
5½ x 8½, sc, 336pp
$29.95

Tribal Bigfoot
David Paulides
978-0-88839-021-9
5½ x 8½,, sc, 480pp
$29.95

Hancock House Publishers
19313 0 Ave, Surrey, BC V3Z 9R9
www.hancockhouse.com
sales@hancockhouse.com
1-800-938-1114

www.ingramcontent.com/pod-product-compliance
Lightning Source LLC
Chambersburg PA
CBHW071417160426
43195CB00013B/1729